Essential Chemistry

Solution Chemistry

4th edition

4 3 2 1

ISBN-13: 979-8-8855706-1-9

Sterling Education materials are available at quantity discounts.

Contact info@sterling–prep.com

Sterling Education
6 Liberty Square #11
Boston, MA 02109

Published by Sterling Education

 Printed in the U.S.A.

STERLING
Education

From the foundations of chemical reactions to the complex mechanisms of atomic particles, *Essential Chemistry Self-Teaching Guides* are a comprehensive compendium of clearly explained texts to learn and master these multifaceted chemistry topics.

These guides provide a detailed review of the fundamental mechanisms of chemical and physical processes at the atomic level. Develop a better understanding of the electronic structure of elements, principles of chemical bonding, phases of matter, types and mechanisms of chemical reactions, and principles of solutions and acid-base equilibria. Learn about rate processes in chemical reactions, empirical and molecular formulas, enthalpy, entropy, oxidation number, the laws of thermodynamics, and electrochemistry. Reinforce your learning by working through the practice questions and step-by-step solutions.

Created by highly qualified chemistry instructors, researchers, and education specialists, these books empower readers by helping them increase their understanding of general chemistry.

We sincerely hope that these guides are valuable for your learning.

250909akp

Electronic Structure & Periodic Table

Chemical Bonding

States of Matter & Phase Equilibria

Stoichiometry

Solution Chemistry

Chemical Kinetics & Equilibrium

Acids & Bases

Chemical Thermodynamics

Electrochemistry

Visit our Amazon store

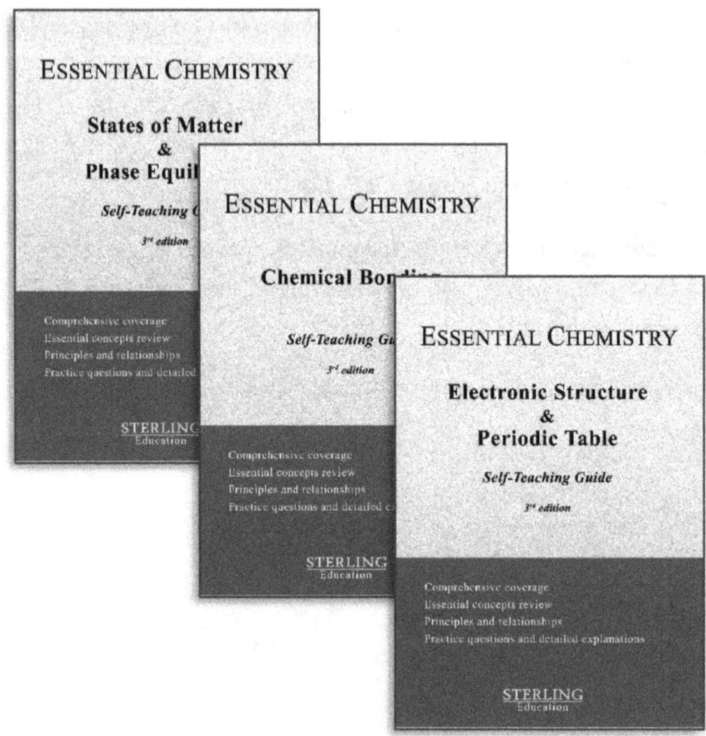

Essential Biology Self-Teaching Guides

Eukaryotic Cell & Cellular Metabolism

Molecular Biology & Genetics

Nervous & Endocrine Systems

Circulatory, Respiratory & Immune Systems

Digestive & Excretory Systems

Muscle, Skeletal & Integumentary Systems

Reproduction & Development

Microbiology

Plants & Photosynthesis

Evolution, Classification & Diversity

Ecology & Population Biology

Visit our Amazon store

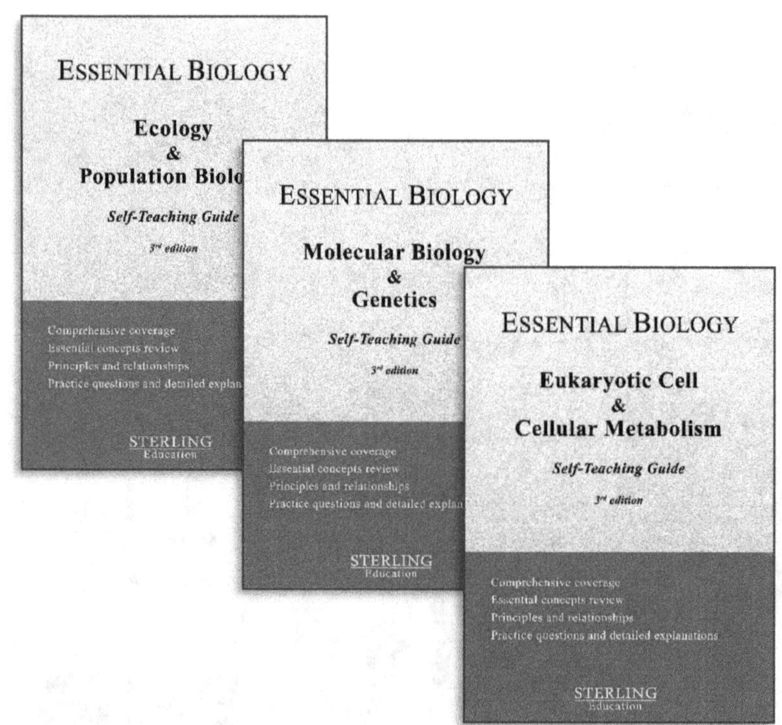

Table of Contents

Table of Contents (*continued*)

REVIEW

Solution Chemistry

Page intentionally left blank

The Nature of Science

Scientific discovery

Observation is the first step toward scientific discovery. Observations recognize patterns and anomalies in unexplained phenomena.

Hypothesis is advanced as a proposed explanation for the observation.

Evaluating a hypothesis involves experiments and additional observations (i.e., recording and collecting data).

Experimental observations are made while manipulating *independent variables* (e.g., how long the plant is exposed to sunlight per twenty-four hours) with *dependent variables* (data collected regarding height measurements).

Observations generate *data* that:

      1) *support the hypothesis* (i.e., different from proves) or

      2) *refute the hypothesis*.

Hypothesis are modified by narrowing or expanding their application to explain additional data. More experiments are performed, and the results either *support* (not prove) or *refute* the hypothesis during this iterative process.

Principles are the initially proposed explanations that are specific and apply to a narrow range of phenomena.

Theories build upon valid hypotheses

Theories are proposed to account for the observations when a hypothesis is valid over a range (e.g., sunlight, temperature, or humidity).

Theories are advanced to explain the phenomena and to predict what will happen under other conditions with different variables.

Further observations determine whether the predictions were accurate.

Research (re-search or *again search*) continues as these many observations support hypotheses and theories.

When theories are accurate over various scenarios and experiments, *Laws* are developed based on these repeatedly replicated results.

Laws as established theories

Laws are brief descriptions of how nature behaves in a broad set of circumstances. They are the consolidation of several theories. Laws are the products of multiple theories tested and proven, culminating in a broad claim regarding those predictions.

Laws are widely applicable explanations. Laws with broad and straightforward claims are more robust than those that make claims on a narrower, more specific set of variables and parameters.

The number of Laws describing physical phenomena is minuscule compared to the number of theories and magnitudes, which are less than the number of supported hypotheses.

The number of refuted (unsupported) hypotheses may be several magnitudes larger.

Failed experiments do not typically refer to the physical process where the researcher made a human error during the experiment. A failed experiment means that the *data does not support the hypothesis.* Therefore, after additional trials to validate the "failed" results, the hypothesis must be modified to predict the results of either future experiments or objective observations in the physical world.

Theories, models and laws

Theories emerge from a hypothesis that experimental data has repeatedly supported. Theories are detailed statements that provide testable predictions of the behavior of natural phenomena.

Theories are established to explain observations and then tested (supported or refuted) based on their predictions. Tested theories can articulate the boundaries of the tested hypothesis.

Theories are more specific than *hypotheses* but vaguer and more encompassing than *models.*

Models are constructed to explain the observed behavior if a hypothesis accurately predicts a phenomenon. They are more specific in their application than theories.

Data collected to develop the theories provide the conceptual foundation for constructing models. The model creates mental pictures (e.g., complicated weather patterns displayed as visual pictures for the viewer) of the physical phenomena.

Models should be consistent with scientific theories' predictions and support a comprehensive understanding of a phenomenon. Models allow for predictive outcomes under specific theories (e.g., when the wind speed reaches x, then y will occur).

Multiple models are integrated to explain a portion, or the entire scope, of a *theory.*

Understand the limitations of a model and do not apply it dogmatically unless its applicability is evaluated. A model is an underlying representation of a part of a theory; it is not intended to provide a complete picture of every phenomenon under that theory.

Instead, models provide the fundamental aspects of the theory.

Solutions

Solutions as homogeneous mixtures

Solutions are homogeneous mixtures of one or more solutes and a solvent, where the components are indistinguishable.

Solution formation involves disrupting the *crystalline lattice structure* of the solute and mixing solute particles with solvent molecules.

Solvent is the substance present in the largest quantity.

Solute particles diffuse into the solution and become uniformly dispersed.

Aqueous solutions have water as the solvent.

Enthalpy considerations for solutions

There must be sufficiently strong interactions between solute particles and solvent molecules for solute particles to diffuse into the solution.

Interactions between solute and solvent molecules must overcome the attractive intermolecular forces between solute particles in the liquid or solid.

Solution processes may be divided into three main stages:

1. pure solvent $\rightarrow$ separated solvent molecules (endothermic);
    $$\Delta H_1 > 0$$

2. pure solute $\rightarrow$ separated solute particles (endothermic);
    $$\Delta H_2 > 0$$

3. separated solvent and solute molecules $\rightarrow$ solution (exothermic);
    $$\Delta H_3 < 0$$

For example:

Solute(s) + Solvent $\rightarrow$ Solution

$$\Delta H_{soln} = \Delta H_1 + \Delta H_2 + \Delta H_3$$

Depending on the magnitude of ΔH_1, ΔH_2 and ΔH_3, the solution process is:

Exothermic: if $|\Delta H_3| > |\Delta H_1 + \Delta H_2|$

Endothermic: if $|\Delta H_3| < |\Delta H_1 + \Delta H_2|$

For solutions, remember the phrase "like dissolves like."

Polar solutes dissolve in polar solvents, while nonpolar solutes dissolve in nonpolar solvents.

Notes for active learning

Hydration energy

Many ionic compounds dissolve in polar water with negative enthalpies (releasing energy) due to *hydration energy.*

Solvation is the attraction of a solvent with solute molecules or ions.

Hydration is the process by which water molecules attract solute molecules or ions.

Hydration energy produced from ion-dipole interactions provides energy to overcome the lattice energy and disrupt ions in crystalline solids (e.g., NaCl).

Water molecules are polar and interact strongly with ionic cations and anions through ion-dipole interactions.

The overall solute–solution formation process may be exothermic or endothermic, depending on whether the solute–solute or solute–solution enthalpy is larger.

Ionic compounds dissolve in water and dissociate (i.e., ionize) to produce free ions (cations and anions).

For example, NaCl dissolves, producing Na^+ and Cl^- ions that become hydrated with water molecules.

Sodium chloride and many other ionic compounds dissolve in water because of the strong ion-dipole interactions between the charged solute particles and the polar solvent (e.g., water) molecules.

For example, for NaCl, the *lattice energy* (i.e., ionic solid to gaseous ions) is slightly higher than the sum of *hydration energy* for Na^+ and Cl^- ions. The solution process for NaCl is slightly endothermic.

Electrolytes

Strong electrolytes (solutions of ionic compounds) dissociate entirely, and the dissolved ions are good conductors of electric current.

Weak electrolytes (solutions of polar covalent compounds) partially dissociate in solution.

Nonelectrolytes do not dissociate into ions (e.g., sugar as a nonelectrolyte) and dissolve in the solution as the original molecules rather than separate into cations and anions.

Anions and cations

Anions are *negatively charged* species, denoted by a single negative sign, indicating a charge of -1.

Cations are *positively charged* species, denoted by a single positive sign, indicating a charge of $+1$.

Charge is written as a superscript.

Example ions include:

ammonium (NH_4^+)

phosphate (PO_4^{3-})

sulfate (SO_4^{2-})

Common anions and cations	Formula
Anions	
Hydroxide	OH^-
Chloride	Cl^-
Hypochlorite	ClO^-
Chlorite	ClO_2^-
Chlorate	ClO_3^-
Perchlorate	ClO_4^-
Halide, hypohalite, etc.	X^-, XO^-, etc.
Carbonate	CO_3^{2-}
Hydrogen Carbonate (Bicarbonate)	HCO_3^-
Sulfate	SO_4^{2-}
Hydrogen Sulfate (Bisulfate)	HSO_4^-
Sulfite	SO_3^{2-}
Thiosulfate	$S_2O_3^{2-}$
Nitrate	NO_3^-
Nitrite	NO_2^-
Phosphate	PO_4^{3-}

Hydrogen Phosphate	HPO_4^{2-}
Dihydrogen Phosphate	$H_2PO_4^-$
Phosphite	PO_3^{3-}
Cyanide	CN^-
Thiocyanate	SCN^-
Peroxide	O_2^{2-}
Oxalate	$C_2O_4^{2-}$
Acetate	$C_2H_3O_2^-$
Chromate	CrO_4^{2-}
Dichromate	$Cr_2O_7^{2-}$
Permanganate	MnO_4^-
Cations	
Hydronium	H_3O^+
Ammonium	NH_4^+
Metal	M^{n+}

Common ions, associated formulas, and charges shown

Identifying ions experimentally

Qualitative analysis is a laboratory method for separating and identifying ions in a mixture.

The technique utilizes solubility differences for ionic compounds in an aqueous solution, as well as a particular cation's ability to form *complex ions* with ligands.

The approach in qualitative analysis of cations separates them sequentially into *ion groups*.

Complex ions of many transition metals are often *colored* and used in visual identity.

Ion group 1: Insoluble chlorides

Treating the mixture with 6 M HCl precipitates Ag^+, Hg_2^{2+}, and Pb^{2+} as chlorides, with other cations in solution.

White precipitate formation indicates at least one of these cations in the mixture.

Ion group 2: Acid-insoluble sulfides

Supernatant from the above treatment with HCl is adjusted to a pH of approximately 0.5 and then treated with aqueous H_2S.

High $[H_3O^+]$ in solution keeps $[HS^-]$ low, which precipitates cation groups: Cu^{2+}, Cd^{2+}, Hg^{2+}, Sn^{2+}, and Bi^{3+}.

Centrifuging and decanting give the next solution.

Ion group 3: Base-insoluble sulfides

Supernatant from acidic sulfide treatment is treated with an NH_3/NH_4^+ buffer to make the solution slightly basic (pH ≈ 8).

Excess OH^- in solution increases $[HS^-]$, which causes precipitation of more soluble sulfides and hydroxides.

Cations that precipitate under this condition of base-insoluble sulfides are Zn^{2+}, Mn^{2+}, Ni^{2+}, Fe^{2+}, and Co^{2+} as sulfides, and Al^{3+}, Cr^{3+}, and Fe^{3+} as hydroxides.

Precipitate is centrifuged, and the supernatant is decanted to obtain the next solution.

Ion group 4: Insoluble phosphates

The slightly basic supernatant separated from the group 3 ions is treated with $(NH_4)_2HPO_4$, which precipitates $Mg_3(PO_4)_2$, $Ca_3(PO_4)_2$ and $Ba_3(PO_4)_2$.

Ion group 5: Alkali metal and ammonium ions

Final solution contains any of the following ions: Na^+, K^+, and NH_4^+.

Solvation and hydration

Solvation is the attraction of a solvent with solute molecules or ions.

Hydration is the attraction of water molecules to solute ions, forming a shell around them in a solution.

For example, the oxygen atom in water is partially negative, so it attracts cations.

Conversely, the hydrogen atoms in water are partially positively charged, surrounding anions.

Hydronium ions

H^+ does not exist as a bare proton in water; it exists as the *hydronium ion* (H_3O^+).

High proton charge density attracts it to a nearby molecule with a partial or full negative charge.

Water has two lone pairs of electrons on the oxygen atom; the positive proton interacts with one of the lone pairs, forming a hydronium ion.

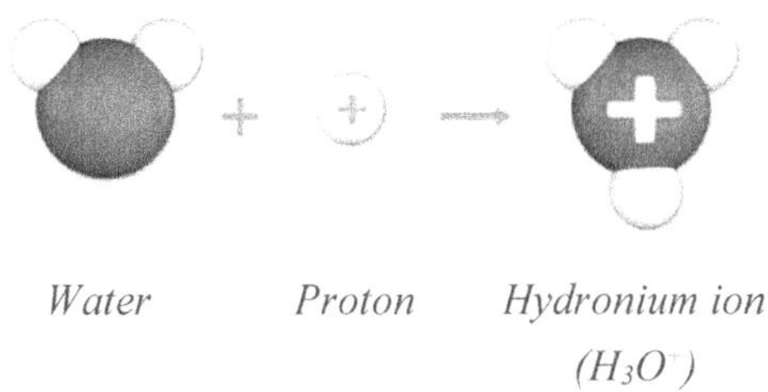

Water   Proton   Hydronium ion

(H_3O^+)

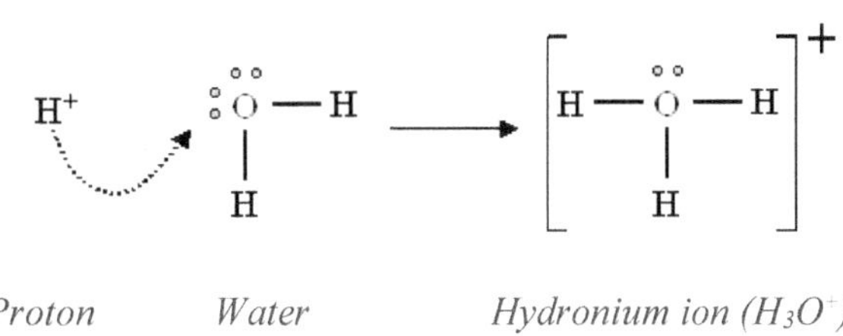

Proton   Water   Hydronium ion (H_3O^+)

Notes for active learning

Solubility Concepts

Solubility

Solubility of a substance is the amount of solute (e.g., grams) that dissolves in a solvent to give a saturated solution at a specific temperature.

Solubility is *temperature-dependent*.

Saturated solutions contain the maximum dissolved solute possible at a given temperature, with a state of dynamic equilibrium existing between dissolution and crystallization.

For example, the solubility of KNO_3 is ~30 g per 100 g of water at 20 °C and 63 g at 40 °C.

Solubility rules for ionic compounds

Use the following rules to *predict the solubility of ionic compounds* in water. These rules generally relate to the compound's lattice and hydration energies of the individual ions.

One factor determining the *lattice energy* is the magnitude of charges on the ions (i.e., the greater the magnitude of the charges, the greater the lattice energy). Simplified, the greater the lattice energy, the lower the solubility.

Rules focus on the magnitudes of the charges and ignore influences on the lattice energies, such as hydration energy. Additionally, the rules overlook the impact of temperature.

Solubility rules are listed in order of *decreasing importance*. For example, if the first rule applies, disregard the subsequent rules. Similarly, if the second rule is applicable, disregard the third rule. Thus, only use the third rule when rules one and two are not applicable.

The following ions are usually soluble: $C_2H_3O_2^-$, ClO_3^-, ClO_4^-. K^+, Na^+, NH_4^+, and NO_3^-.

General principle: Hydrides (H^-) decompose in water to yield H_2 and ^-OH. Many metal oxides react with water to produce hydroxides (^-OH).

Rule 1: compounds containing a +1 or –1 ion are typically soluble

Examples include: $AlCl_3$, $FeCl_2$, K_2SO_4, $NaBr$, Rb_3PO_4

Exceptions (Insoluble): OH^- (other than with cations that produce strong bases)

Hg_2^{2+} (other than with NO_3^-, $C_2H_3O_2^-$, ClO_3^-, and ClO_4^-)

Pb^{2+}, Hg^{2+} with most –1 ions

IB metals (column 11)

Rule 2: compounds containing a +3 or –3 or higher ion are typically NOT soluble

Examples include: $AlPO_4$, $Ca_3(PO_4)_2$, TiO2

Exceptions (Soluble):   cations combined with sulfate or dichromate

Rule 3: compounds containing –2 ions are typically NOT soluble

Examples include: $CaCO_3$, FeS, ZnC_2O_4

Exceptions (soluble):

  Dichromates and sulfates (insoluble if combined with Ba, Ca, Hg, Pb, and Sr)

There are exceptions to the solubility rules besides those listed. However, the ions commonly encountered follow these rules.

Solubility examples for ionic compounds

AgBr	Insoluble	Rule 1	Ag^+
AgCl	Insoluble	Rule 1	$Ag^+ + Cl^-$
$AgIO_3$	Insoluble	Rule 1	$Ag^+ + IO_3^-$
AgOH	Insoluble	Rule 1	weak base
Ag_2S	Insoluble	Rule 1	Ag^+
Hg_2Cl_2	Insoluble	Rule 1	Hg_2^{2+}
$PbCl_2$	Insoluble	Rule 1	Pb^{2+}
$AgNO_3$	Soluble	Rule 1	$NO_3^- + Ag^+$
$Ba(NO_3)_2$	Soluble	Rule 1	NO_3^-
$Ba(OH)_2$	Soluble	Rule 1	strong base
NaCl	Soluble	Rule 1	$Na^+ + Cl^-$
$NaClO_3$	Soluble	Rule 1	$Na^+ + ClO_3^-$
NaOH	Soluble	Rule 1	strong base
Na_3PO_4	Soluble	Rule 1	Na^+

$(NH_4)_2CO_3$	Soluble	Rule 1	NH_4^+
NH_4IO_3	Soluble	Rule 1	$NH_4^+ + IO_3^-$
$FeAsO_4$	Insoluble	Rule 2	$Fe^{3+} + AsO_4^{3-}$
$Fe_2(SO_4)_3$	Soluble	Rule 2	SO_4^{2-}
$BaCO_3$	Insoluble	Rule 3	CO_3^{2-}
$BaSO_4$	Insoluble	Rule 3	$Ba^{2-} + SO_4^{2-}$
$MgSO_4$	Insoluble	Rule 3	SO_4^{2-}
$PbCrO_4$	Insoluble	Rule 3	CrO_4^{2-}
ZnS	Insoluble	Rule 3	S^{2-}
$Ba(IO_3)_2$	Insoluble	exception	

Solution concentration

When a solution that is almost saturated at a higher temperature is cooled slowly where solubility is lower, excess solutes typically precipitate, resulting in a *saturated solution* at the lower temperature.

However, if the solution is cooled rapidly, precipitates do not form, and the resulting solution contains more dissolved solute than it would in a typical saturated solution. This process produces a *supersaturated solution*.

> *Supersaturated solutions* are unstable, and crystallization occurs through seeding (i.e., introducing particles that provide nuclei for precipitation).

> *Concentrated solutions* contain a relatively large number of dissolved solutes, typically containing 40 g or more of dissolved solute in 100 mL of water.

> *Dilute solutions* contain a large amount of solvent and a small number of dissolved solutes, with less than 10 g of dissolved solute per 100 mL of water.

Concentrated solution does *not* necessarily imply a *saturated* solution.

Saturated solution does *not* necessarily imply a *concentrated* solution.

Temperature and pressure affect solubility

Solubility of most liquids and solids in water *increases* with temperature.

Gas solubility *decreases* with temperature.

Pressure of a gas strongly affects its solubility, as described by *Henry's Law*, which states that solubility increases as pressure increases (i.e., directly proportional).

Factors affecting solubility include solute's *surface area* (e.g., granulated sugar dissolves faster than a sugar cube) and *heating or agitating a solution* (e.g., stirring accelerates dissolving).

Miscible and *immiscible* describe liquids that dissolve or do not dissolve in another liquid, respectively. For example, ethanol and water are *miscible* because they dissolve in each other freely when mixed.

Oil and water are *immiscible* since they remain separate.

Units of concentration

Solution concentration is expressed for the solute amount dissolved and expressed as:

$$\text{Mass Percent, \% (w/w)} = \frac{\text{mass of solute}}{\text{mass of solution}} \times 100\%$$

$$\text{Volume Percent \% (v/v)} = \frac{\text{volume of solute}}{\text{volume of solution}} \times 100\%$$

$$\text{Molarity (M)} = \frac{\text{number of moles of solute}}{\text{liters of solution}}$$

$$\text{Molality (m)} = \frac{\text{number of moles of solute}}{\text{kg of solvent}}$$

$$\text{Mole fraction, } (X_i) = \frac{\text{moles of a component}}{\text{total moles of components in the solution}}$$

$$\text{Normality } (N) = \frac{\text{number of equivalents of solute}}{\text{liters of solution}}$$

where

$$\text{Number of equivalents of solute} = \frac{\text{grams of solute}}{\text{equivalent weight of solute}}$$

Normality and equivalent weight

Normality is based on the chemical mass unit of equivalent weight.

Equivalent weight is the amount of solute needed to equal one mole of hydrogen ions.

Equivalent weight is thus dependent on the *valence of the solute.*

For solutes with a valence of 1 (e.g., HCl), the molecular and equivalent weights are the same.

For solutes with a valence greater than 1 (e.g., H_3PO_4, valence equals 3), the equivalent weight equals the *molecular weight divided by the valence.*

Therefore, molarity is related to normality, as in the examples:

$$1 \text{ M HCl} = 1 \text{ N HCl}$$

$$1 \text{ M } H_2SO_4 = 2 \text{ N } H_2SO_4$$

$$1 \text{ M } H_3PO_4 = 3 \text{ N } H_3PO_4$$

Since ionic compounds dissociate into ions, the total number of particles in a solution of an ionic compound is greater than in a solution of a nonionic (molecular) compound.

Total concentration of ions in the solution is the sum of the concentrations of cations and anions, depending on the compound's formula.

For example, a 1.0 M aluminum nitrate solution, $Al(NO_3)_3$, contains 1.0 M of Al^{3+} and 3.0 M NO_3^- ions, with the total concentration of ions in the solution of 4.0 M.

	Before dissolving		*After dissolving*	
$Al(NO_3)_3\,(aq)$	$\rightarrow$	$Al^{3+}\,(aq)$	$+$	$3\,NO_3^-\,(aq)$
(1.0 M)	$\rightarrow$	(1.0 M)		$(3 \times 1.0 \text{ M})$

For example, a solution of 1.0 M $MgCl_2$ contains 1.0 M of Mg^{2+} and 2.0 M of Cl^- ions, and the total ion concentration is 3.0 M.

Notes for active learning

Solution Equilibrium

Solubility product constant (K_{sp})

Solubility product constants (K_{sp}) describe saturated solutions of low-solubility ionic compounds.

For example, when a slightly soluble salt such as silver chloride (AgCl) dissolves in water, a saturated solution is obtained rapidly. A minimal number of solid dissolves, while most AgCl salt remains undissolved.

Equilibrium between AgCl (s) and Ag^+ (aq) + Cl^- (aq) ions occurs in solution:

$$AgCl\ (s) \rightleftarrows Ag^+\ (aq) + Cl^-\ (aq)$$

$$K_{sp} = [Ag^+] \cdot [Cl^-]$$

General expression of the solubility product constant (K_{sp}) for solubility equilibrium is:

$$M_aX_b\ (s) \rightleftarrows aM^{b+}\ (aq) + bX^{a-}\ (aq)$$

$$K_{sp} = [M^{b+}]^a \cdot [X^{a-}]^b$$

In some problems, K_{sp} is given, and the solubility needs to be calculated; or the solubility is given, and K_{sp} needs to be calculated.

The procedure to follow depends on the type of equilibrium.

Sample solubility problems are solved below.

Ionic equilibrium

A. Ionic equilibria of the type:

$$MX\ (s) \rightleftarrows M^{n+}\ (aq) + X^{n-}\ (aq)$$

$$K_{sp} = [M^{n+}] \cdot [X^{n-}]$$

If solubility (S) of compounds is S mol/L:

$$K_{sp} = S^2$$

$$S = \sqrt{(K_{sp})}$$

For example, solubility equilibrium (K_{sp}) for $BaSO_4$ is:

$$BaSO_4\,(s) \rightleftarrows Ba^{2+}\,(aq) + SO_4^{2-}\,(aq)$$

$$K_{sp} = [Ba^{2+}]{\cdot}[SO_4^{2-}]$$

$$K_{sp} = 1.5 \times 10^{-9}$$

If the solubility (S) of $BaSO_4$ is S mol/L, a saturated solution of $BaSO_4$ has:

$$[Ba^{2+}] = [SO_4^{2-}] = S \text{ mol/L}$$

$$K_{sp} = S^2$$

$$S = \sqrt{(K_{sp})}$$

$$S = \sqrt{(1.5 \times 10^{-9})}$$

$$S = 3.9 \times 10^{-5} \text{ mol/L}$$

B. Ionic equilibria of the type:

$$MX_2\,(s) \rightleftarrows M^{2+}\,(aq) + 2\,X^-\,(aq)$$

$$K_{sp} = [M^{2+}]{\cdot}[X^-]^2$$

For the type:

$$M_2X\,(s) \rightleftarrows 2M^+\,(aq) + X^{2-}\,(aq)$$

$$K_{sp} = [M^+]^2{\cdot}[X^{2-}]$$

For both types, if solubility is S mol/L:

$$K_{sp} = 4S^3$$

$$S = (K_{sp}/4)^{1/3}$$

For example:

$$CaF_2\,(s) \rightleftarrows Ca^{2+}\,(aq) + 2\,F^-\,(aq)$$

$$K_{sp} = [Ca^{2+}]{\cdot}[F^-]^2$$

$$K_{sp} = 4.0 \times 10^{-11}$$

Solubility (S) of calcium fluoride (CaF_2) is:

$$S = (K_{sp} / 4)^{1/3}$$

$$S = (4.0 \times 10^{-11} / 4)^{1/3}$$

$$S = 2.2 \times 10^{-4} \text{ mol/L}$$

C. Solubility equilibria of the type:

$$MX_3\,(s) \rightleftarrows M^{3+}\,(aq) + 3\ X^-\,(aq)$$

$$K_{sp} = [M^{3+}]\cdot[X^-]^3$$

Or of the type:

$$M_3X\,(s) \rightleftarrows 3\ M^+\,(aq) + X^{3-}\,(aq)$$

$$K_{sp} = [M^+]^3\cdot[X^{3-}]$$

If the solubility (S) of the compound (MX_3 or M_3X) is S mol/L:

$$K_{sp} = 27S^4$$

$$S = \sqrt[4]{(K_{sp} / 27)}$$

For example:

$$Ag_3PO_4\,(s) \rightleftarrows 3Ag^+\,(aq) + PO_4^{3-}\,(aq)$$

$$K_{sp} = [Ag^+]^3\cdot[PO_4^{3-}]$$

$$K_{sp} = 1.8 \times 10^{-18}$$

Solubility (S) of silver phosphate (Ag_3PO_4) is:

$$S = \sqrt[4]{(K_{sp} / 27)}$$

$$S = \sqrt[4]{(1.8 \times 10^{-18} / 27)}$$

$$S = 1.6 \times 10^{-5} \text{ mol/L}$$

D. Solubility equilibria of the type:

$$M_2X_3\,(s) \rightleftarrows 2\ M^{3+}\,(aq) + 3\ X^{2-}\,(aq)$$

$$K_{sp} = [M^{3+}]^2\cdot[X^{2-}]^3$$

Or of the type:

$$M_3X_2\,(s) \rightleftarrows 3\,M^{2+}\,(aq) + 2\,X^{3-}\,(aq)$$

$$K_{sp} = [M^{2+}]^3 \cdot [X^{3-}]^2$$

If solubility (S) of the compound M_3X_2 is S mol/L:

$$[M^{2+}] = 3S \text{ and } [X^{3-}] = 2S$$

$$K_{sp} = (3S)^3 \cdot (2S)^2$$

$$K_{sp} = 108S^5$$

$$S = \sqrt[5]{(K_{sp}/108)}$$

For example:

$$Ca_3(PO_4)_2\,(s) \rightleftarrows 3\,Ca^{2+}\,(aq) + 2\,PO_4^{3-}\,(aq)$$

$$K_{sp} = [Ca^{2+}]^3 \cdot [PO_4^{3-}]^2$$

$$K_{sp} = 1.3 \times 10^{-32}$$

Solubility (S) of calcium phosphate ($Ca_3(PO_4)_2$) is:

$$S = 5\sqrt{(1.3 \times 10^{-32})/108}$$

$$S = 1.6 \times 10^{-7} \text{ mol/L}$$

Calculating K_{sp} from solubility

For example, the solubility (S) of $PbSO_4$ in water is 4.3×10^{-3} g/100 mL solution at 25 °C. What is the K_{sp} of $PbSO_4$ at 25 °C?

$$\text{Solubility } (S) \text{ of } PbSO_4 \text{ in mol/L} = \frac{4.3 \times 10^{-3} \text{ g}}{100 \text{ mL}} \times \frac{1000 \text{ mL/L}}{303.26 \text{ g/mol}}$$

$$S = 1.40 \times 10^{-4} \text{ mol/L}$$

Saturated solution of $PbSO_4$ contains:

$$[Pb^{2+}] = [SO_4^{2-}] = 1.4 \times 10^{-4} \text{ mol/L}$$

For the equilibrium:

$$PbSO_4\,(s) \rightleftarrows Pb^{2+}\,(aq) + SO_4^{2-}\,(aq)$$

$$K_{sp} = [Pb^{2+}] \cdot [SO_4^{2-}]$$

$$S^2 = (1.4 \times 10^{-4} \text{ mol/L})^2$$

$$S = 2.0 \times 10^{-8}$$

Calculating solubility from K_{sp}

For example, if K_{sp} of $Mg(OH)_2$ is 6.3×10^{-10} at 25 °C, what is its solubility in mol/L at 25 °C?

Determine the solubility equilibrium for magnesium hydroxide ($Mg(OH)_2$):

$$Mg(OH)_2\,(s) \rightleftharpoons Mg^{2+}\,(aq) + 2\,OH^-\,(aq)$$

$$K_{sp} = [Mg^{2+}]\cdot[OH^-]^2$$

$$4S^3 = 6.3 \times 10^{-10}$$

Solubility of $Mg(OH)_2$:

$$S = \sqrt[3]{(K_{sp}/4)}$$

$$S = \sqrt[3]{(6.3 \times 10^{-10}/4)}$$

$$S = 5.4 \times 10^{-4}\ \text{mol/L}$$

If K_{sp} of $Mg(OH)_2$ is 6.3×10^{-10} at 25 °C, what is its solubility in a g/100 mL solution at 25 °C?

Solubility in g/100 mL solution:

$$S = (5.4 \times 10^{-4}\ \text{mol/L})\cdot(58.32\ \text{g/mol})\cdot(0.1\ \text{L/100 mL})$$

$$S = 3.1 \times 10^{-3}\ \text{g/100 mL solution}$$

Notes for active learning

Common and Complex Ions

Common ions

Common ions are produced by more than one solute in solution.

According to *Le Châtelier's Principle*, the following equilibrium exists for acetic acid:

$$HC_2H_3O_2 \, (aq) + H_2O \, (l) \leftrightarrows H_3O^+ \, (aq) + C_2H_3O_2^- \, (aq)$$

Reaction shifts to the *left* if $C_2H_3O_2^-$ is introduced from another source. The acetate ion ($C_2H_3O_2^-$) is the common ion in sodium acetate and acetic acid solution.

This effect reduces the degree of dissociation of the acid, decreases $[H_3O^+]$, and increases the pH of the solution.

For example, the addition of ammonium chloride, NH_4Cl, causes the reaction to shift left in the equilibrium of ammonia in an aqueous solution:

$$NH_3 \, (aq) + H_2O \, (l) \leftrightarrows NH_4^+ \, (aq) + OH^- \, (aq)$$

NH_4Cl dissociates into NH_4^+ and Cl^-, where NH_4^+ is a common ion in ammonia equilibrium.

The equilibrium shift caused by NH_4^+ ion reduces the extent of ionization of ammonia, which decreases $[OH^-]$ in the system and lowers the pH of the solution.

Common-ion effect

Common-ion effect is Le Châtelier's Principle applied to K_{sp} reactions. It states that a common ion *decreases* the solubility of a slightly soluble ionic compound.

A common ion is an ion in a solution that is common to the ionic compound.

For example, in the following equilibrium:

$$PbCl_2 \, (s) \rightleftharpoons Pb^{2+} \, (aq) + 2 \, Cl^- \, (aq)$$

If $NaCl$ is added to a saturated solution of $PbCl_2$, the $[Cl^-]$ increases, and according to Le Châtelier's Principle, the equilibrium shifts in the direction that reduces $[Cl^-]$.

In this example, the equilibrium shifts to the left, forming more $PbCl_2$ solid, thereby decreasing the amount of dissolved $PbCl_2$. More $PbCl_2$ dissolves in pure water than in water with Cl^- ions.

Common-ion effect *selectively precipitates* a mixture component during laboratory separations.

For example, add $NaCl$ to separate $AgCl$ from a mixture of $AgCl$ and Ag_2SO_4. The addition of $NaCl$ selectively displaces $AgCl$ by the common-ion effect (Cl^- being the common ion).

Complex ion formation

Complex ion consists of a central metal ion covalently bonded to two or more *ligands*, which can be anions such as ^-OH, Cl^-, F^- and ^-CN, or neutral molecules such as H_2O, CO, and NH_3.

Metal ions form complex ions with water molecules as ligands in aqueous solutions.

Ligand exchanges occur when another ligand is introduced into solution, establishing equilibrium.

For example, in the complex ion $[Cu(NH_3)_4]^{2+}$, Cu^{2+} is the central metal ion, with four NH_3 molecules covalently bonded to it.

Complex ions are Lewis adducts (i.e., the addition of a Lewis acid and a Lewis base).

Lewis bases can be charged or uncharged.

Lewis acids and bases

Metal ions act as *Lewis acids* (electron-pair acceptors).

Ligands are *Lewis bases* (electron-pair donors).

$$\text{Metal}^+ + \text{Lewis base} \rightarrow \text{Complex ion}$$

$$M^+ + L \rightarrow M - L_n^{\;+}$$

K_{eq} for this reaction is K_f, or the *formation constant*.

Formation constant K_f

For example, when NH_3 is added to an aqueous solution containing Cu^{2+} ions, the following equilibrium occurs:

$$Cu(H_2O)_6^{2+} \, (aq) \; + \; 4\,NH_3 \, (aq) \rightleftarrows \; [Cu(NH_3)_4]^{2+} \, (aq) + 6\,H_2O$$

$$K_f = [Cu(NH_3)_4^{2+}] \,/\, [Cu(H_2O)_6^{2+}] \cdot [NH_3]^4$$

Ligand exchange occurs stepwise; each water molecule is sequentially replaced with an NH_3 molecule to yield a series of intermediate species, each with its formation constant (K_f).

For convenience, the water molecule is omitted.

1. $Cu^{2+} \, (aq) + NH_3 \, (aq) \rightleftarrows Cu(NH_3)^{2+} \, (aq)$

    $$K_{f1} = [Cu(NH_3)^{2+}] \,/\, [Cu^{2+}] \cdot [NH_3]$$

2. $Cu(NH_3)^{2+} \, (aq) + NH_3 \, (aq) \rightleftarrows Cu(NH_3)_2^{2+} \, (aq)$

    $$K_{f2} = [Cu(NH_3)_2^{2+}] \,/\, [Cu(NH_3)^{2+}] \cdot [NH_3]$$

3. $Cu(NH_3)_2^{2+}\ (aq) + NH_3\ (aq) \rightleftarrows\ Cu(NH_3)_3^{2+}\ (aq)$

$$K_{f3} = [Cu(NH_3)_3^{2+}] / [Cu(NH_3)_2^{2+}] \cdot [NH_3]$$

4. $Cu(NH_3)_3^{2+}\ (aq) + NH_3\ (aq) \rightleftarrows\ Cu(NH_3)_4^{2+}\ (aq)$

$$K_{f4} = [Cu(NH_3)_4^{2+}] / [Cu(NH_3)_3^{2+}] \cdot [NH_3]$$

Formation constant (K_f) is the product of the intermediate formation constants:

$$K_f = K_{f1} \times K_{f2} \times K_{f3} \times K_{f4}$$

$$K_f = [Cu(NH_3)_4^{2+}] / [Cu^{2+}] \cdot [NH_3]^4$$

Solubility of complex ions

Complex-ion effect is the opposite of the *common*-ion effect.

Ligands *increase* solubility of slightly soluble ionic compounds if complex ions form with metal ions.

For example, silver chloride (AgCl) is *more soluble* in ammonia (NH$_3$) solution because silver ions form complex ions with NH$_3$:

$$AgCl\ (s) \rightleftarrows Ag^+\ (aq) + Cl^-\ (aq)$$

$$K_{sp} = 1.6 \times 10^{-10}$$

$$Ag^+\ (aq) + 2\ NH_3\ (aq) \rightleftarrows Ag(NH_3)_2^+\ (aq)$$

$$K_f = 1.7 \times 10^7$$

$$AgCl\ (s) + 2\ NH_3\ (aq) \rightleftarrows Ag(NH_3)_2^+\ (aq) + Cl^-\ (aq)$$

$$K_{net} = K_{sp} \times K_f$$

$$K_{net} = (1.6 \times 10^{-10}) \cdot (1.7 \times 10^7)$$

$$K_{net} = 2.7 \times 10^{-3}$$

Cl^- ion is reduced when a complex ion forms, so more AgCl dissolves.

Alternatively:

$$AgCl\ (s) \leftrightarrow Ag^+\ (aq) + Cl^-\ (aq)$$

$$NH_3 + Ag^+ \leftrightarrow Ag\text{-}(NH_3)_n\ \text{complex ion}$$

Complex ion formation reduces Ag^+, and more AgCl dissolves.

pH and solubility

pH affects the solubility of slightly soluble compounds containing anions as *conjugate bases of weak acids* (e.g., F^-, NO_2^-, OH^-, SO_3^{2-} and PO_4^{3-}).

pH *does not* affect the solubility of slightly soluble compounds containing anions as *conjugate bases of strong acids* (SO_4^{2-}, Cl^- and Br^-).

In a saturated solution of calcium fluoride (CaF_2) equilibrium exists:

$$CaF_2\ (s) \rightleftarrows Ca^{2+}\ (aq) + 2\ F^-\ (aq)$$

If a strong acid is added to the saturated solution, the following reaction occurs:

$$H^+\ (aq) + F^-\ (aq) \rightarrow HF\ (aq)$$

The reaction has the net effect of *reducing* F^- ion concentration, causing the equilibrium to shift to the *right*, and more CaF_2 dissolves.

Acids are more soluble in bases.

$$HA \rightarrow H^+ + A^-$$

Placing the above reactions in a base reduces the H^+.

Thus, more HA dissolves, according to Le Châtelier's Principle.

Bases are more soluble in acids.

$$B + H^+ \rightarrow BH^+$$

Putting the above reactions in an acid adds H^+, and thus, more B dissolves, according to Le Châtelier's Principle.

Notes for active learning

Notes for active learning

Practice Questions
&
Detailed Explanations

Practice Questions

Practice Set 1: Questions 1–20

1. If the solubility of nitrogen in blood is 1.90 cc/100 cc at 1.00 atm, what is the solubility of nitrogen in a scuba diver's blood at a depth of 125 feet where the pressure is 4.5 atm?

   A. 1.90 cc/100 cc

   B. 2.36 cc/100 cc

   C. 4.5 cc/100 cc

   D. 0.236 cc/100 cc

   E. 8.55 cc/100 cc

2. All of the statements about molarity are correct, EXCEPT:

   A. volume = moles/molarity

   B. moles = molarity × volume

   C. molarity of a diluted solution is less than the molarity of the original solution

   D. abbreviation is M

   E. molarity equals moles of solute per mole of solvent

3. Which of the following molecules is expected to be most soluble in water?

   A. $NaCl$

   B. $CH_3CH_2CH_2COOH$

   C. $CH_3CH_2CH_2\,OH$

   D. $Al(OH)_3$

   E. CH_4

4. The equation for the reaction shown below can be written as an ionic equation.

$$BaCl_2\ (aq) + K_2CrO_4\ (aq) \rightarrow BaCrO_4\ (s) + 2\ KCl\ (aq)$$

In the ionic equation, the spectator ions are:

   A. K^+ and Cl^-

   B. Ba^{2+} and CrO_4^{2-}

   C. Ba^{2+} and K^+

   D. K^+ and CrO_4^{2-}

   E. Cl^- and CrO_4^{2-}

5. Carbon dioxide gas is injected into soda in commercially prepared soft drinks. Under what conditions are carbon dioxide gas most soluble?

    A. High temperature, high-pressure

    B. High temperature, low-pressure

    C. Low temperature, low-pressure

    D. Low temperature, high-pressure

    E. Solubility is the same for all conditions

6. A solute is a:

    A. substance that dissolves into a solvent

    B. substance containing a solid, liquid or gas

    C. solid substance that does not dissolve into water

    D. solid substance that does not dissolve at a given temperature

    E. liquid that does not dissolve into another liquid

7. How many ions are produced in solution by dissociating one formula unit of $Co(NO_3)_2 \cdot 6H_2O$?

    A. 2   **B.** 3   **C.** 4   **D.** 6   **E.** 9

8. 15 grams of an unknown substance is dissolved in 60 grams of water. When the solution is transferred to another container, it weighs 78 grams. Which of the following is a possible explanation?

    A. The solution reacted with the second container, forming a precipitate

    B. Some of the solution remained in the first container

    C. The reaction was endothermic, which increased the average molecular speed

    D. The solution reacted with the first container, causing some byproducts to be transferred with the solution

    E. The reaction was exothermic, which increased the average molecular speed

9. Which of the following is NOT soluble in H_2O?

    A. Iron (III) hydroxide   **C.** Potassium sulfate

    B. Iron (III) nitrate   **D.** Ammonium sulfate

                                  E. Sodium chloride

10. What is the v/v% concentration of a solution made by adding 25 mL of acetone to 75 mL of water?

A. 33% v/v

B. 0.33% v/v

C. 25% v/v

D. 2.5% v/v

E. 3.3% v/v

11. Why is octane less soluble in H_2O than in benzene?

A. Bonds between benzene and octane are much stronger than between H_2O and octane

B. Octane cannot dissociate in the presence of H_2O

C. Bonds between H_2O and octane are weaker than the bonds between H_2O molecules

D. Octane and benzene have similar molecular weights

E. H_2O dissociates in the presence of octane

12. What is the K_{sp} for slightly soluble copper (II) phosphate in an aqueous solution?

$$Cu_3(PO_4)_2 \; (s) \leftrightarrow 3 \; Cu^{2+} \; (aq) + 2 \; PO_4{}^{3-} \; (aq)$$

A. $K_{sp} = [Cu^{2+}]^3 \cdot [PO_4{}^{3-}]^2$

B. $K_{sp} = [Cu^{2+}] \cdot [PO_4{}^{3-}]^2$

C. $K_{sp} = [Cu^{2+}]^3 \cdot [PO_4{}^{3-}]$

D. $K_{sp} = [Cu^{2+}] \cdot [PO_4{}^{3-}]$

E. $K_{sp} = [Cu^{2+}]^2 \cdot [PO_4{}^{3-}]^3$

13. Apply the *like dissolves like* rule to predict which of the following liquids is/are miscible with water:

    I.   carbon tetrachloride, CCl_4

    II.   toluene, C_7H_8

    III.   ethanol, C_2H_5OH

A. I only

B. II only

C. III only

D. I and II only

E. I, II and III

14. In which of the following pairs of substances would both species in the pair be written in the molecular form in a net ionic equation?

    I. CO_2 and H_2SO_4 II. LiOH and H_2 III. HF and CO_2

A. I only

B. II only

C. III only

D. I, II and III

E. None of the above

15. Which statement best describes a supersaturated solution?

    A. It contains dissolved solute in equilibrium with undissolved solid

    B. It rapidly precipitates if a seed crystal is added

    C. It contains as much solvent as it can accommodate

    D. It contains no double bonds

    E. It contains only electrolytes

16. What is the mass of a 7.50% urine sample that contains 122 g of dissolved solute?

    A. 1,250 g   **B.** 935 g   **C.** 49.35 g   **D.** 155.4 g   **E.** 1,627 g

17. Calculate the molarity of a solution prepared by dissolving 15.0 g of NH_3 in 250 g of water with a final density of 0.974 g/mL.

    A. 36.2 M   **C.** 0.0462 M

    B. 3.23 M   **D.** 0.664 M

                                              E. 6.80 M

18. Which of the following compounds has the highest boiling point?

    A. 0.2 M $Al(NO_3)_3$ **C.** 0.2 M glucose $(C_6H_{12}O_6)$

    B. 0.2 M $MgCl_2$ **D.** 0.2 M Na_2SO_4

                                            E. Pure H_2O

19. Which compound produces four ions per formula unit by dissociation when dissolved in water?

    A. Li_3PO_4 **C.** $MgSO_4$

    B. $Ca(NO_3)_2$ **D.** $(NH_4)_2SO_4$

                                            E. $(NH_4)_4Fe(CN)_6$

20. Which of the following would be a weak electrolyte in a solution?

    A. HBr (*aq*)   **C.** KOH

    B. KCl   **D.** $HC_2H_3O_2$

                                          E. HI

Practice Set 2: Questions 21–40

21. The ions Ca^{2+}, Mg^{2+}, Fe^{2+}, Fe^{3+}, present in groundwater, can be removed by pretreating the water with:

A. $PbSO_4$

B. $Na_2CO_3 \cdot 10H_2O$

C. KNO_3

D. $CaCl_2$

E. 0.05 M HCl

22. Choose the spectator ions: $Pb(NO_3)_2$ *(aq)* + H_2SO_4 *(aq)* $\rightarrow$?

A. NO_3^- and H^+

B. H^+ and SO_4^{2-}

C. Pb^{2+} and H^+

D. Pb^{2+} and NO_3^-

E. Pb^{2+} and SO_4^{2-}

23. From the *like dissolves like* rule, predict which of the following vitamins is soluble in water:

A. α-tocopherol ($C_{29}H_{50}O_2$)

B. calciferol ($C_{27}H_{44}O$)

C. ascorbic acid ($C_6H_8O_6$)

D. retinol ($C_{20}H_{30}O$)

E. none of the above

24. How much water must be added when 125 mL of a 2.00 M solution of HCl is diluted to a final concentration of 0.400 M?

A. 150 mL

B. 850 mL

C. 625 mL

D. 750 mL

E. 500 mL

25. Which of the following is the sulfate ion?

A. SO_4^{2-}

B. S^{2-}

C. CO_3^{2-}

D. PO_4^{3-}

E. S^-

26. Which of the following explains why bubbles form inside a pot of water when the pot of water is heated?

    A. As temperature increases, the vapor pressure increases

    B. As temperature increases, the atmospheric pressure decreases

    C. As temperature increases, the solubility of air decreases

    D. As temperature increases, the kinetic energy decreases

    E. None of the above

27. A solution in which the rate of crystallization is equal to the rate of dissolution is:

    A. saturated   **C.** dilute

    B. supersaturated   **D.** unsaturated

                                       E. impossible to determine

28. What is the term that refers to liquids that do not dissolve in one another and separate into two layers?

    A. Soluble   **C.** Insoluble

    B. Miscible   **D.** Immiscible

                                       E. None of the above

29. Which is a correctly balanced hydration equation for the hydration of Na_2SO_4?

    A. $Na_2SO_4 \ (s) \xrightarrow{H_2O} Na^+ \ (aq) + 2SO_4^{2-} \ (aq)$

    B. $Na_2SO_4 \ (s) \xrightarrow{H_2O} 2\,Na^{2+} \ (aq) + S^{2-} \ (aq) + O_4^{2-} \ (aq)$

    C. $Na_2SO_4 \ (s) \xrightarrow{H_2O} Na_2^{2+} \ (aq) + SO_4^{2-} \ (aq)$

    D. $Na_2SO_4 \ (s) \xrightarrow{H_2O} 2\,Na^+ \ (aq) + SO_4^{2-} \ (aq)$

    E. $Na_2SO_4 \ (s) \xrightarrow{H_2O} 2\,Na^{2+} \ (aq) + S^{2-} \ (aq) + SO_4^{2-} \ (aq) + O_4^{2-} \ (aq)$

30. Soft drinks are carbonated by injection with carbon dioxide gas. Under what conditions is carbon dioxide gas least soluble?

A. High temperature, low-pressure

B. High temperature, high-pressure

C. Low temperature, high-pressure

D. Low temperature, low-pressure

E. None of the above

31. Which type of compound is likely to dissolve in H_2O?

    I.   One with hydrogen bonds

    II. Highly polar compound

    III. Salt

A. I only

B. II only

C. III only

D. I and III only

E. I, II and III

32. Which of the following might have the best solubility in water?

A. CH_3CH_3

B. CH_3OH

C. CCl_4

D. O_2

E. None of the above

33. What is the K_{sp} for slightly soluble gold (III) chloride in an aqueous solution for the reaction shown?

$$AuCl_3\,(s) \leftrightarrow Au^{3+}\,(aq) + 3\ Cl^-\,(aq)$$

A. $K_{sp} = [Au^{3+}]^3{\cdot}[Cl^-] / [AuCl_3]$

B. $K_{sp} = [Au^{3+}]{\cdot}[Cl^-]^3$

C. $K_{sp} = [Au^{3+}]^3{\cdot}[Cl^-]$

D. $K_{sp} = [Au^{3+}]{\cdot}[Cl^-]$

E. $K_{sp} = [Au^{3+}]{\cdot}[Cl^-]^3 / [AuCl_3]$

34. What is the net ionic equation for the reaction shown?

$$CaCO_3 + 2\ HNO_3 \rightarrow Ca(NO_3)_2 + CO_2 + H_2O$$

A. $CO_3^{2-} + H^+ \rightarrow CO_2$

B. $CaCO_3 + 2\ H^+ \rightarrow Ca^{2+} + CO_2 + H_2O$

C. $Ca^{2+} + 2\ NO_3^- \rightarrow Ca(NO_3)_2$

D. $CaCO_3 + 2\ NO_3^- \rightarrow Ca(NO_3)_2 + CO_3^{2-}$

E. None of the above

35. What is the volume of a 0.550 M $Fe(NO_3)_3$ solution needed to supply 0.950 moles of nitrate ions?

A. 265 mL

B. 0.828 mL

C. 22.2 mL

D. 576 mL

E. 384 mL

36. What is the molarity of a solution that contains 48 mEq Ca^{2+} per liter?

A. 0.024 M

B. 0.048 M

C. 1.8 M

D. 2.4 M

E. 0.96 M

37. If 36.0 g of LiOH is dissolved in water to make 975 mL of solution, what is the molarity of the LiOH solution? (Use molecular mass of LiOH = 24.0 g/mol)

A. 1.54 M

B. 2.48 M

C. 0. 844 M

D. 0.268 M

E. 0.229 M

38. What volume of 8.50% (m/v) solution contains 60.0 grams of glucose?

A. 170 mL

B. 448 mL

C. 706 mL

D. 344 mL

E. 960 mL

39. In an AgCl solution, if the K_{sp} for AgCl is A, and the concentration Cl^- in a container is B molar, what is the concentration of Ag (in moles/liter)?

I. A moles/liter

II. B moles/liter

III. A/B moles/liter

A. I only

B. II only

C. III only

D. II and III only

E. I and III only

40. Which statement below is generally true?

    A. Bases are strong electrolytes and ionize completely when dissolved in water

    B. Salts are strong electrolytes and dissociate completely when dissolved in water

    C. Acids are strong electrolytes and ionize completely when dissolved in water

    D. Bases are weak electrolytes and ionize completely when dissolved in water

    E. Salts are weak electrolytes and ionize partially when dissolved in water

Notes for active learning

Practice Set 3: Questions 41–60

41. Which of the following intermolecular attractions is/are important for the formation of a solution?

        I. solute-solute

        II. solvent-solute

        III. solvent-solvent

A. I only

B. III only

C. I and II only

D. II and III only

E. I, II and III

42. What is the concentration of I^- ions in a 0.40 M solution of magnesium iodide?

A. 0.05 M

B. 0.80 M

C. 0.60 M

D. 0.20 M

E. 0.40 M

43. Which is most likely soluble in NH_3?

A. CO_2

B. SO_2

C. CCl_4

D. N_2

E. H_2

44. Which of the following represents the symbol for the chlorite ion?

A. ClO_2^- **B.** ClO^- **C.** ClO_4^- **D.** ClO_3^- **E.** ClO_2

45. Which statement best describes what is happening in a water softening unit?

A. Sodium is removed from the water, making water interact less with the soap molecules

B. Ions in the water softener are softened by chemically bonding with sodium

C. Hard ions are trapped in the softener, which filters out the ions

D. Hard ions in water are exchanged for ions that do not interact as strongly with soaps

E. None of the above

46. Which of the following compounds are soluble in water?

      I. $Mn(OH)_2$ II. $Cr(NO_3)_3$ III. $Ni_3(PO_4)_2$

A. I only

B. II only

C. III only

D. I and III only

E. I, II and III

47. The hydration number of an ion is the number of:

A. water molecules bonded to an ion in an aqueous solution

B. water molecules required to dissolve one mole of ions

C. ions bonded to one mole of water molecules

D. ions dissolved in one liter of an aqueous solution

E. water molecules required to dissolve the compound

48. When a solid dissolves, each molecule is removed from the crystal by interaction with the solvent. This process of surrounding each ion with solvent molecules is called:

A. hemolysis

B. electrolysis

C. crenation

D. dilution

E. solvation

49. The term *miscible* describes which type of solution?

A. Solid/solid

B. Liquid/gas

C. Liquid/solid

D. Liquid/liquid

E. Solid/gas

50. Apply the *like dissolves like* rule to predict which of the following vitamins is insoluble in water:

A. niacinamide ($C_6H_6N_2O$)

B. pyridoxine ($C_8H_{11}NO_3$)

C. retinol ($C_{20}H_{30}O$)

D. thiamine ($C_{12}H_{17}N_4OS$)

E. cyanocobalamin ($C_{63}H_{88}CoN_{14}O_{14}P$)

51. Which species is NOT written as constituent ions when the equation is expanded into the ionic equation?

$$Mg(OH)_2\ (s) + 2\ HCl\ (aq) \rightarrow MgCl_2\ (aq) + 2\ H_2O\ (l)$$

A. $Mg(OH)_2$ only

B. H_2O and $Mg(OH)_2$

C. HCl

D. $MgCl_2$

E. HCl and $MgCl_2$

52. If x moles of $PbCl_2$ fully dissociate in 1 liter of H_2O, the K_{sp} is equivalent to:

A. x^2 **B.** $2x^4$ **C.** $3x^2$ **D.** $2x^3$ **E.** $4x^3$

53. Which of the following are strong electrolytes?

I. salts   II. strong bases   III. weak acids

A. I only

B. I and II only

C. III only

D. I, II and III

E. I and III only

54. What are the spectator ions in the reaction between KOH and HNO_3?

A. K^+ and NO_3^-

B. H^+ and NO_3^-

C. K^+ and H^+

D. H^+ and ^-OH

E. K^+ and ^-OH

55. What volume of 14 M acid is diluted with distilled water to prepare 6.0 L of 0.20 M acid?

A. 86 mL

B. 62 mL

C. 0.94 mL

D. 6.8 mL

E. 120 mL

56. What is the molarity of a solution when diluting 160 mL of 4.50 M NaOH to 595 mL?

A. 0.242 M

B. 1.21 M

C. 2.42 M

D. 1.72 M

E. 0.115 M

57. Which compound is most likely to be more soluble in the nonpolar solvent of benzene than in water?

    A. SO_2

    B. CO_2

    C. Silver chloride

    D. H_2S

    E. CH_2Cl_2

58. Which of the following concentrations is dependent on temperature?

    A. Mole fraction

    B. Molarity

    C. Mass percent

    D. Molality

    E. More than one of the above

59. Which of the following solutions is the most concentrated?

    A. One liter of water with 1 gram of sugar

    B. One liter of water with 2 grams of sugar

    C. One liter of water with 5 grams of sugar

    D. One liter of water with 10 grams of sugar

    E. All are the same

60. What mass of NaOH is contained in 75.0 mL of a 5.0% (w/v) NaOH solution?

    A. 6.50 g

    B. 15.0 g

    C. 7.50 g

    D. 0.65 g

    E. 3.75 g

Practice Set 4: Questions 61–80

Questions **61** through **63** are based on the following data:

	K_{sp}
$PbCl_2$	1.0×10^{-5}
$AgCl$	1.0×10^{-10}
$PbCO_3$	1.0×10^{-15}

61. Consider a saturated solution of $PbCl_2$. The addition of NaCl would:

    I. decrease $[Pb^{2+}]$
    II. increase the precipitation of $PbCl_2$
    III. do not affect the precipitation of $PbCl_2$

A. I only

B. II only

C. III only

D. I and II only

E. I and III only

62. What occurs when $AgNO_3$ is added to a saturated solution of $PbCl_2$?

    I. AgCl precipitates
    II. $Pb(NO_3)_2$ forms a white precipitate
    III. More $PbCl_2$ forms

A. I only

B. II only

C. III only

D. I, II and III

E. I and II only

63. Comparing equal volumes of saturated solutions for $PbCl_2$ and AgCl, which solution contains a greater concentration of Cl^-?

    I. $PbCl_2$
    II. AgCl
    III. Both have the same concentration of Cl^-

A. I only

B. II only

C. III only

D. I and II only

E. Cannot be determined

64. Which of the following is why hexane is significantly soluble in octane?

A. Entropy increases for the two substances as the dominant factor in the ΔG when mixed

B. Hexane hydrogen bonds with octane

C. Intermolecular bonds between hexane-octane are much stronger than either hexane or octane molecular bonds

D. ΔH for hexane-octane is greater than hexane-H_2O

E. Hexane and octane have similar molecular weights

65. Which of the following are characteristics of an ideally dilute solution?

    I. Solute molecules do not interact with each other

    II. Solvent molecules do not interact with each other

    III. The mole fraction of the solvent approaches 1

A. I only

B. II only

C. I and III only

D. I, II and III

E. I and II only

66. Which principle states that the solubility of a gas in a liquid is proportional to the partial pressure of the gas above the liquid?

A. Solubility principle

B. Tyndall effect

C. Colloid principle

D. Henry's law

E. None of the above

67. Water and methanol are two liquids that dissolve in each other. When the two are mixed, they form one layer because the liquids are:

A. unsaturated

B. saturated

C. miscible

D. immiscible

E. supersaturated

68. What is the molarity of a glucose solution that contains 10.0 g of $C_6H_{12}O_6$ dissolved in 100.0 mL of solution? (Use the molecular mass of $C_6H_{12}O_6 = 180.0$ g/mol)

A. 1.80 M

B. 0.555 M

C. 0.0555 M

D. 0.00555 M

E. 18.0 M

69. Which of the following solid compounds is insoluble in water?

    I. $BaSO_4$ II. Hg_2Cl_2 III. $PbCl_2$

A. I only

B. II only

C. III only

D. I and III only

E. I, II and III

70. The net ionic equation for the reaction between zinc and the hydrochloric acid solution is:

A. $Zn\ (s) + 2\ H^+\ (aq) + 2\ Cl^-\ (aq) \rightarrow Zn^{2+}\ (aq) + 2\ Cl^-\ (aq) + H_2\ (g)$

B. $ZnCl_2\ (aq) + H_2\ (g) \rightarrow Zn\ (s) + 2\ HCl\ (aq)$

C. $Zn\ (s) + 2\ H^+\ (aq) \rightarrow Zn^{2+}\ (aq) + H_2\ (g)$

D. $Zn\ (s) + 2\ HCl\ (aq) \rightarrow ZnCl_2\ (aq) + H_2\ (g)$

E. None of the above

71. Apply the *like dissolves like* rule to predict which of the following liquids are miscible with water:

    I.   methyl ethyl ketone, C_4H_8O

    II. glycerin, $C_3H_5(OH)_3$

    III. formic acid, $HCHO_2$

A. I only

B. II only

C. III only

D. I and II only

E. I, II and III

72. What is the K_{sp} for calcium fluoride (CaF_2) if the calcium ion concentration in a saturated solution is 0.00021 *M*?

A. $K_{sp} = 3.7 \times 10^{-11}$

B. $K_{sp} = 2.6 \times 10^{-10}$

C. $K_{sp} = 3.6 \times 10^{-9}$

D. $K_{sp} = 8.1 \times 10^{-10}$

E. $K_{sp} = 7.3 \times 10^{-11}$

73. A 4 M solution of H_3A is completely dissociated in water. How many equivalents of H^+ are found in 1/3 liter?

A. ¼   **B.** 1   **C.** 1.5   **D.** 3   **E.** 4

74. Which of the following aqueous solutions are poor conductors of electricity?

    I.   sucrose, $C_{12}H_{22}O_{11}$

    II. barium nitrate, $Ba(NO_3)_2$

    III. calcium bromide, $CaBr_2$

A. I only

B. II only

C. III only

D. I and II only

E. I, II and III

75. Which of the following solid compounds is insoluble in water?

A. $BaSO_4$

B. Na_2S

C. $(NH_4)_2CO_3$

D. K_2CrO_4

E. $Sr(OH)_2$

76. Which is true if the ion concentration product of a solution of AgCl is less than the K_{sp}?

    I.   Precipitation occurs

    II. The ions are insoluble in water

    III. Precipitation does not occur

A. I only

B. II only

C. III only

D. I and II only

E. II and III only

77. Under which conditions is the expected solubility of oxygen gas in water the highest?

A. High temperature and high O_2 pressure above the solution

B. Low temperature and low O_2 pressure above the solution

C. Low temperature and high O_2 pressure above the solution

D. High temperature and low O_2 pressure above the solution

E. The O_2 solubility is independent of temperature and pressure

78. What is the molarity of an 8.60 molal solution of methanol (CH_3OH) with a 0.94 g/mL density?

A. 0.155 M

B. 23.5 M

C. 6.34 M

D. 9.68 M

E. 2.35 M

79. Which of the following is NOT a unit factor related to a 15.0% aqueous solution of potassium iodide (KI)?

A. 100 g solution / 85.0 g water

B. 85.0 g water / 15.0 g KI

C. 85.0 g water / 100 g solution

D. 15.0 g KI / 85.0 g water

E. 15.0 g KI / 100 g water

80. What is the molar concentration of a solution containing 0.75 mol of solute in 75 cm^3 of solution?

A. 0.1 M   **B.** 1.5 M   **C.** 3 M   **D.** 10 M   **E.** 1 M

Notes for active learning

Practice Set 5: Questions 81–100

81. If 25.0 mL of seawater has a mass of 25.88 g and contains 1.35 g of solute, what is the mass/mass percent concentration of solute in the seawater sample?

A. 1.14%

B. 12.84%

C. 2.62%

D. 5.22%

E. 9.45%

82. Which of the following statements describing solutions is NOT true?

A. Solutions are colorless

B. The particles in a solution are atomic or molecular

C. Making a solution involves a physical change

D. Solutions are homogeneous

E. Solutions are transparent

83. Calculate the solubility product of AgCl if the solubility of AgCl in H_2O is 1.3×10^{-4} mol/L?

A. 1.3×10^{-4}

B. 1.3×10^{-2}

C. 2.6×10^{-4}

D. 3.9×10^{-5}

E. 1.7×10^{-8}

84. Which of the following solutions is the most dilute?

A. 0.1 liters of H_2O with 1 gram of sugar

B. 0.2 liters of H_2O with 2 grams of sugar

C. 0.5 liters of H_2O with 5 grams of sugar

D. 1 liter of H_2O with 10 grams of sugar

E. All have the same concentration

85. When salt A is dissolved into water to form a 1 molar unsaturated solution, the temperature of the solution decreases. Under these conditions, which statement is accurate when salt A is dissolved in water?

A. $\Delta H°$ and $\Delta G°$ are positive

B. $\Delta H°$ is positive and $\Delta G°$ is negative

C. $\Delta H°$ is negative and $\Delta G°$ is positive

D. $\Delta H°$ and $\Delta G°$ are negative

E. $\Delta H°$, $\Delta S°$ and $\Delta G°$ are positive

86. The heat of a solution measures the energy absorbed during:

    I.   the formation of solvent-solute bonds

    II. the breaking of solute-solute bonds

    III. the breaking of solvent-solvent bonds

A. I only

B. II only

C. I and II only

D. I and III only

E. II and III only

87. Why does the reaction proceed if, when solid potassium chloride is dissolved in H_2O, the energy of the bonds formed is less than the energy of the bonds broken?

A. The electronegativity of the H_2O increases by interactions with potassium and chloride ions

B. The reaction does not take place under standard conditions

C. The decreased disorder due to mixing decreases entropy within the system

D. Remaining potassium chloride which does not dissolve offsets the portion that dissolves

E. The increased disorder due to mixing increases entropy within the system

88. Why are salts more soluble in H_2O than in benzene?

A. Benzene is aromatic and therefore very stable

B. The dipole moment of H_2O compensates for the loss of ionic bonding when salt dissolves

C. Strong intermolecular attractions in benzene must be disrupted to dissolve salt in benzene

D. The molecular mass of H_2O is similar to the atomic mass of most ions

E. The dipole moment of H_2O compensates for the increased ionic bonding when salt dissolves

89. What is the formula of the solid formed when aqueous barium chloride is mixed with aqueous potassium chromate?

A. K_2CrO_4

B. K_2Ba

C. KCl

D. $BaCrO_4$

E. $BaCl_2$

90. Why might sodium carbonate (washing soda, Na_2CO_3) be added to hard water for cleaning?

A. The soap gets softer due to the added ions

B. The ions solubilize the soap due to ion-ion intermolecular attraction, which improves the cleaning ability

C. The hard ions in the water are more attracted to the carbonate ions' –2 charge

D. The added sodium ions dissolve the hard ions

E. None of the above

91. What volume of 0.25 M hydrochloric acid reacts completely with 0.400 g of sodium hydrogen carbonate, $NaHCO_3$? (Use the molar mass of $NaHCO_3$ = 84.0 g/mol)

$$NaHCO_3 \ (s) + HCl \ (aq) \rightarrow NaCl \ (aq) + H_2O \ (l) + CO_2 \ (g)$$

A. 142 mL

B. 34.6 mL

C. 19.0 mL

D. 54.6 mL

E. 32.6 mL

92. What is the concentration of the contaminant in ppm (m/m) if 8.8×10^{-3} g of a contaminant is present in 5,246 g of a particular solution?

A. 12.40 ppm

B. 1.53 ppm

C. 4.06 ppm

D. 1.68 ppm

E. 8.72 ppm

93. Which of the following structures represents the bicarbonate ion?

A. HCO_3^-

B. $H_2CO_3^{2-}$

C. CO_3^-

D. CO_3^{2-}

E. HCO_3^{2-}

94. Of the following, which can serve as the solute in a solution?

I. solid   II. liquid   III. gas

A. I only
B. II only

C. III only
D. I and II only
E. I, II and III

95. Hydration involves the:

A. formation of water–solute bonds
B. breaking of water–water bonds
C. breaking of water–solute bonds
D. breaking of solute–solute bonds
E. breaking of water-water bonds and formation of water-solute bonds

96. Which is the name for a substance represented by a formula written as $M_xLO_y \cdot zH_2O$?

A. Solvent
B. Solute

C. Solid hydrate
D. Colloid
E. Suspension

97. In the reaction between aqueous silver nitrate and aqueous potassium chromate, what is the identity of the soluble substance that is formed?

A. Potassium nitrate
B. Potassium chromate

C. Silver nitrate
D. Silver chromate
E. No soluble substance is formed

98. Apply the *like dissolves like* rule to predict which liquid is miscible with liquid bromine (Br_2).

I.   benzene (C_6H_6)
II. hexane (C_6H_{14})
III. carbon tetrachloride (CCl_4)

A. I only
B. II only

C. III only
D. I and II only
E. I, II and III

99. Hydrochloric acid contains:

A. ionic bonds and is a weak electrolyte

B. hydrogen bonds and is a weak electrolyte

C. covalent bonds and is a strong electrolyte

D. ionic bonds and is a non-electrolyte

E. covalent bonds and is a non-electrolyte

100. At 22 °C, a one-liter sample of pure water has a vapor pressure of 18.5 torr. If 7.5 g of NaCl is added to the sample, what is the result for the vapor pressure of the water?

A. It equals the vapor pressure of the NaCl added to the sample

B. It remains unchanged at 18.5 torr

C. It increases the vapor pressure

D. It decreases the vapor pressure

E. It increases to the square root of solute added

Notes or active learning

Answer Key & Detailed Explanations

Answer Key

1: E	11: C	21: B	31: E	41: E	51: B	61: D	71: E	81: D	91: C
2: E	12: A	22: A	32: B	42: B	52: E	62: A	72: A	82: A	92: D
3: A	13: C	23: C	33: B	43: B	53: B	63: A	73: E	83: E	93: A
4: A	14: C	24: E	34: B	44: A	54: A	64: A	74: A	84: E	94: E
5: D	15: B	25: A	35: D	45: D	55: A	65: C	75: A	85: B	95: E
6: A	16: E	26: C	36: A	46: B	56: B	66: D	76: C	86: E	96: C
7: B	17: B	27: A	37: A	47: A	57: B	67: C	77: C	87: E	97: A
8: D	18: A	28: D	38: C	48: E	58: B	68: B	78: C	88: B	98: E
9: A	19: A	29: D	39: D	49: D	59: D	69: E	79: E	89: D	99: C
10: C	20: D	30: A	40: B	50: C	60: E	70: C	80: D	90: C	100: D

===

Practice Set 1: Questions 1–20

===

1. E is correct.

Solubility is proportional to pressure.

Use simple proportions to compare solubility in different pressures:

$$P_1 / S_1 = P_2 / S_2$$

$$S_2 = P_2 / (P_1 / S_1)$$

$$S_2 = 4.5 \text{ atm} / (1.00 \text{ atm} / 1.90 \text{ cc}/100 \text{ mL})$$

$$S_2 = 8.55 \text{ cc} / 100 \text{ mL}$$

2. E is correct.

The correct interpretation of molarity is *moles of solute per liter of solvent.*

3. A is correct.

NaCl is a charged, ionic salt (Na^+ and Cl^-) and is extremely water-soluble.

Hexanol hydrogen bonds with water due to the hydroxyl group, but the long hydrocarbon chain reduces its solubility, and the chains interact via hydrophobic interactions forming micelles.

Aluminum hydroxide has poor solubility in water and requires the addition of Brønsted-Lowry acids to dissolve completely.

4. A is correct.

Spectator ions appear on both sides of the ionic equation.

Ionic equation of the reaction:

$$Ba^{2+}\,(aq) + 2\ Cl^-\,(aq) + 2\ K^+\,(aq) + CrO_4^-\,(aq) \rightarrow BaCrO_4\,(s) + 2\ K^+\,(aq) + 2\ Cl^-\,(aq)$$

5. D is correct.

Higher pressure exerts more pressure on the solution, allowing more molecules to dissolve in the solvent.

Lower temperature increases gas solubility.

As the temperature increases, solvent and solute molecules move faster, and it is more difficult for the solvent molecules to bond with the solute.

6. A is correct.

Solutes dissolve in the solvent to form a solution.

7. B is correct.

When the compound is dissolved in water, the attached hydrate/water crystals dissociate and become part of the water solvent.

Therefore, the only ions left are 1 Co^{2+} and 2 NO_3^-.

8. D is correct.

The mass of a solution is calculated by taking the mass of the solution and container and subtracting the mass of the container.

9. A is correct.

An insoluble compound cannot be dissolved (especially with reference to water).

Hydroxide salts of Group I elements are soluble, while hydroxide salts of Group II elements (Ca, Sr and Ba) are slightly soluble.

Salts containing nitrate ions (NO_3^-) are generally soluble.

Most sulfate salts are soluble; important exceptions: $BaSO_4$, $PbSO_4$, Ag_2SO_4 and $SrSO_4$.

Hydroxide salts of transition metals and Al^{3+} are insoluble.

Thus, $Fe(OH)_3$, $Al(OH)_3$ and $Co(OH)_2$ are not soluble.

10. C is correct.

% v/v solution = (volume of acetone / volume of solution) × 100%

% v/v solution = {25 mL / (25 mL + 75 mL)} × 100%

% v/v solution = 25%

11. C is correct.

The van der Waals forces in the hydrocarbons are relatively weak, while hydrogen bonds between H_2O molecules are strong intermolecular bonds.

Bonds between polar H_2O and nonpolar octane are weaker than those between two polar H_2O molecules (i.e., hydrogen bonds).

12. A is correct.

For K_{sp}, take the molarities (or concentrations) of the products (cC and dD) and multiply them.

For K_{sp}, only aqueous species are included in the calculation.

If any of the products have coefficients, raise the product to that coefficient power and multiply the concentration by that coefficient:

$$K_{sp} = [C]^c \cdot [D]^d$$

$$K_{sp} = [Cu^{2+}]^3 \cdot [PO_4^{3-}]^2$$

The reactant (aA) is solid and is not included in the K_{sp} equation.

Solids are *not* included when calculating equilibrium constant expressions because their concentrations do not change the expression. Any change in their concentrations is insignificant and is thus omitted.

13. C is correct.

The *like dissolves like* rule applies when a solvent is miscible with a solute that has similar properties.

A polar solute (e.g., ethanol) is miscible with a polar solvent (e.g., water).

14. C is correct.

In a net ionic equation, substances that do not dissociate in aqueous solutions are written in their molecular form (not broken down into ions).

Gases, liquids, and solids are written in molecular form.

Some solutions do not dissociate into ions in water or do so in very little amounts (e.g., weak acids such as HF, CH_3COOH). They are also written in molecular form.

15. B is correct.

A *supersaturated* solution contains more of the dissolved material than could be dissolved by the solvent under normal conditions.

Increased heat allows for a solution to become supersaturated.

The term also refers to the vapor of a compound with a higher partial pressure than the vapor pressure of that compound.

A supersaturated solution forms a precipitate if seed crystals are added as the solution reaches a lower energy state when solutes precipitate from the solution.

16. E is correct.

Mass % = mass of solute / mass of solution

Rearrange that equation to solve for mass of solution:

mass of solution = mass of solute / mass %

mass of solution = 122 g / 7.50%

mass of solution = 122 g / 0.075

mass of solution = 1,627 g

17. B is correct.

Determine moles of NH_3:

Moles of NH_3 = mass of NH_3 / molecular weight of NH_3

Moles of NH_3 = 15.0 g / (14.01 g/mol + 3 × 1.01 g/mol)

Moles of NH_3 = 15.0 g / (17.04 g/mol)

Moles of NH_3 = 0.88 mol

Determine the volume of the solution (solvent + solute):

Volume of solution = mass / density

Volume of solution = (250 g + 15 g) / 0.974 g/mL

Volume of solution = 272.1 mL

Convert volume to liters:

Volume = 272.1 mL × 0.001 L / mL

Volume = 0.2721 L

Divide moles by the volume to calculate molarity:

Molarity = moles / liter of solution

Molarity = 0.88 mol / 0.2721 L = 3.23 M

18. A is correct.

Solutions with the highest concentration of ions have the highest boiling point.

Calculate the concentration of ions in each solution:

0.2 M $Al(NO_3)_3$ = 0. 2 M × 4 ions = 0.8 M

0. 2 M $MgCl_2$ = 0. 2 M × 3 ions = 0.6 M

0. 2 M glucose = 0. 2 M × 1 ion = 0.2 M (glucose does not dissociate into ions in solution)

0. 2 M Na_2SO_4 = 0. 2 M × 3 ions = 0.6 M

Water = 0 M

19. A is correct.

Calculate the number of ions in each option:

$$A: Li_3PO_4 \rightarrow 3\,Li^+ + PO_4^{3-} \qquad (4\ ions)$$

$$B: Ca(NO_3)_2 \rightarrow Ca^{2+} + 2\,NO_3^- \qquad (3\ ions)$$

$$C: MgSO_4 \rightarrow Mg^{2+} + SO_4^{2-} \qquad (2\ ions)$$

$$D: (NH_4)_2SO_4 \rightarrow 2\,NH_4^+ + SO_4^{2-} \qquad (3\ ions)$$

$$E: (NH_4)_4Fe(CN)_6 \rightarrow 4\,NH_4^+ + Fe(CN)_6^{4-} \qquad (5\ ions)$$

20. D is correct.

An *electrolyte* is a substance that produces an electrically conducting solution when dissolved in a polar solvent (e.g., water).

The dissolved electrolyte separates into positively-charged cations and negatively-charged anions.

Strong electrolytes dissociate entirely (or almost completely) because the resulting ions are stable in the solution.

==

Practice Set 2: Questions 21–40

==

21. B is correct.

Note that carbonate salts help remove 'hardness' in water.

Generally, SO_4, NO_3 and Cl salts tend to be soluble in water, while CO_3 salts are less soluble.

22. A is correct.

A *spectator ion* exists in the same form as reactants and products of a chemical reaction.

Balanced equation:

$$Pb(NO_3)_2\ (aq) + H_2SO_4\ (aq) \rightarrow PbSO_4\ (s) + 2\ HNO_3\ (aq)$$

23. C is correct.

A polar solute is miscible with a polar solvent.

Ascorbic acid is polar and is therefore miscible in water.

24. E is correct.

Apply the formula for the molar concentration of a solution:

$$M_1V_1 = M_2V_2$$

Substitute the given volume and molar concentrations of HCl, solve for the final volume of HCl of the resulting dilution:

$$V_2 = [M_1V_1] / (M_2)$$

$$V_2 = [(2.00\ M\ HCl) \times (0.125\ L\ HCl)] / (0.400\ M\ HCl)$$

$$V_2 = 0.625\ L\ HCl$$

Solve for the volume of water needed to be added to the initial volume of HCl to obtain the final diluted volume of HCl:

$$V_{H_2O} = V_2 - V_1$$

$$V_{H_2O} = (0.625\ L\ HCl) - (0.125\ L\ HCl)$$

$$V_{H_2O} = 0.500\ L = 500\ mL$$

25. A is correct.

The *–ate* ending indicates the species with more oxygen than species ending in *–ite*.

However, it does not indicate a specific number of oxygen molecules.

26. C is correct.

Solubility is the property of a solid, liquid, or gaseous substance (i.e., solute); it dissolves in a solid, liquid, or gaseous solvent to form a solution (i.e., solute in the solvent).

The solubility of a substance depends on the physical and chemical properties of the solute and solvent and the temperature, pressure, and pH of the solution.

Gaseous solutes (e.g., oxygen) exhibit complex behavior with temperature. As the temperature rises, gases usually become less soluble in water but more soluble in organic solvents.

27. A is correct.

A *saturated* solution contains the maximum dissolved material in the solvent under normal conditions. The term refers to the vapor of a compound with a higher partial pressure than the vapor pressure of that compound.

Increased heat allows for a solution to become supersaturated.

A saturated solution forms a precipitate as more solute is added to the solution.

28. D is correct.

Immiscible refers to the property (of solutions) of when two or more substances (e.g., oil and water) are mixed and eventually separate into two layers.

Miscible is when two liquids are mixed but do not necessarily interact chemically.

In contrast to miscibility, *soluble* means the substance (solid, liquid, or gas) can *dissolve* in another solid, liquid, or gas.

In other words, a substance *dissolves* when it becomes incorporated into another substance.

In contrast to miscibility, solubility involves a *saturation point* at which a substance cannot dissolve any further and a mass, the *precipitate*, begins to form.

29. D is correct.

In this case, hydration means dissolving in water rather than reacting with water. Therefore, the compound dissociates into its ions (without involving water in the actual chemical reaction).

30. A is correct.

Depending on the solubility of a solute, there are three possible results:

> 1) a dilute solution has less solute than the maximum amount able to dissolve;

> 2) a saturated solution has precisely the same amount as its solubility;

> 3) a precipitate forms if there is more solute than can be dissolved, so the excess solute separates from the solution (i.e., crystallization).

Precipitation lowers the solute concentration to the saturation level to increase the stability of the solution.

Gas solubility is inversely proportional to temperature and proportional to pressure.

31. E is correct.

Like dissolves like means that polar substances tend to dissolve in polar solvents and nonpolar substances in nonpolar solvents.

Molecules that can form hydrogen bonds with water are soluble.

Salts are ionic compounds.

The anion and cation bond with the polar molecule of water and are soluble.

32. B is correct.

Like dissolves like means that polar substances tend to dissolve in polar solvents and nonpolar substances in nonpolar solvents.

Methanol (CH_3OH) is a polar molecule.

Therefore, methanol is soluble in water because it *hydrogen bonds* with water.

33. B is correct.

To find the K_{sp}, take the molarities or concentrations of the products (cC and dD) and multiply them. For K_{sp}, only aqueous species are included in the calculation.

If any of the products have coefficients, raise the product to that coefficient power and multiply the concentration by that coefficient:

$$K_{sp} = [C]^c \cdot [D]^d$$

$$K_{sp} = [Au^{3+}] \cdot [Cl^-]^3$$

The reactant (aA) is solid and is not included in the K_{sp} equation.

Solids are not included when calculating equilibrium constant expressions because their concentrations do not change the expression.

Any change in their concentrations is insignificant and is thus omitted.

34. B is correct.

Break down the molecules into their constituent ions:

$$CaCO_3 + 2\ H^+ + 2\ NO_3^- \rightarrow Ca^{2+} + 2\ NO_3^- + CO_2 + H_2O$$

Remove species that appear on both sides of the reaction:

$$CaCO_3 + 2\ H^+ \rightarrow Ca^{2+} + CO_2 + H_2O$$

35. D is correct.

According to the problem, there are 0.950 moles of nitrate ion in a $Fe(NO_3)_3$ solution.

Because there are three nitrates (NO_3) ions for each $Fe(NO_3)_3$ molecule, the moles of $Fe(NO_3)_3$ can be calculated:

$$(1\ mole\ /\ 3\ mole) \times 0.950\ moles = 0.317\ moles$$

Calculate the volume of solution:

Volume of solution = moles of solute / molarity

Volume of solution = 0.317 moles / 0.550 mol/L

Volume of solution = 0.576 L

Convert the volume into milliliters: $0.576\ L \times 1{,}000\ mL/L = 576\ mL$

36. A is correct.

mEq represents the amount in milligrams of a solute equal to 1/1,000 of its gram equivalent weight, considering the valency of the ion.

Millimolar (mM) = mEq / valence.

Divide the given concentration of Ca^{2+} by 2:

$$[Ca^{2+}] = [48 \text{ mEq } Ca^{2+}] / 2 = 24 \text{ mM } Ca^{2+}$$

$$[Ca^{2+}] = 24 \text{ mM } Ca^{2+}$$

Divide the concentration of Ca^{2+} by 1,000 to obtain the concentration of Ca^{2+} in units of molarity:

$$[Ca^{2+}] = 24 \text{ mM } Ca^{2+} \times [(1 \text{ M}) / (1,000 \text{ mM})]$$

$$[Ca^{2+}] = 0.024 \text{ M } Ca^{2+}$$

37. A is correct.

Start by calculating the number of moles:

Moles of LiOH = mass of LiOH / molar mass of LiOH

Moles of LiOH = 36.0 g / (24.0 g/mol)

Moles of LiOH = 1.50 moles

Divide moles by volume to calculate molarity:

Molarity = moles / volume

Molarity = 1.50 moles / (975 mL × 0.001 L/mL)

Molarity = 1.54 M

38. C is correct.

The glucose content is 8.50 % (m/v); the solute (glucose) is measured in grams, but the solution volume is measured in milliliters.

Therefore, the mass and volume of the solution are interchangeable (1 g = 1 mL).

% (m/v) of glucose = mass of glucose / volume of solution

volume of solution = mass of glucose / % (m/v) of glucose

volume of solution = 60 g / 8.50%

volume of solution = 706 mL

39. D is correct.

When AgCl dissociates, equal amounts of Ag^+ and Cl^- are produced.

If the concentration of Cl^- is B, this must be the concentration of Ag^+.

Therefore, B can be the concentration of Ag.

The concentration of silver ion can be determined by dividing the K_{sp} by the concentration of chloride ion:

$$K_{sp} = [Ag^+] \cdot [Cl^-].$$

Therefore, the concentration of Ag can be A/B moles/liter.

40. B is correct.

Strong electrolytes dissociate entirely (or almost completely) in water.

Strong acids and bases dissociate almost completely, but weak acids and bases dissociate only slightly.

Practice Set 3: Questions 41–60

41. E is correct.

Solute-solute and solvent-solvent attractions are important in establishing bonding amongst themselves, both for solvent and solute.

Once they are mixed in a solution, the solute-solvent attraction becomes the major attraction force, but the other two forces (solute-solute and solvent-solvent attractions) are still present.

42. B is correct.

Since the empirical formula for magnesium iodide is MgI_2, two moles of dissolved I^- result from each mole of dissolved MgI_2.

Therefore, if $[MgI_2] = 0.40$ M, then $[I^-] = 2(0.40$ M$) = 0.80$ M.

43. B is correct.

Ammonia forms hydrogen bonds, and SO_2 is polar.

44. A is correct.

The chlorite ion (chlorine dioxide anion) is ClO_2^-.

Chlorite is a compound that contains this group, with chlorine in an oxidation state of +3.

formula	Cl^-	ClO^-	ClO_2^-	ClO_3^-	ClO_4^-
Anion name	chloride	hypochlorite	chlorite	chlorate	perchlorate
Oxidation state	−1	+1	+3	+5	+7

45. D is correct.

Water softeners cannot remove ions from the water without replacing them.

The process replaces the ions that cause scaling (precipitation) with non-reactive ions.

46. B is correct.

These compounds are rarely encountered in most chemistry problems and are part of the long list of exceptions to the solubility rule.

Salts containing nitrate ions (NO_3^-) are generally soluble.

47. A is correct.

Hydration involves the interaction of water molecules with the solute. The water molecules exchange bonding relationships with the solute, whereby water-water bonds break and water-solute bonds form.

When an ion is hydrated, it is surrounded and bonded by water molecules. The average number of water molecules bonding to an ion is its *hydration number*.

Hydration numbers can vary but often are either 4 or 6.

48. E is correct.

Solvation describes the process whereby the solvent surrounds the solute molecules.

49. D is correct.

Miscible refers to the property (of solutions) when two or more substances (e.g., water and alcohol) are mixed without separating.

50. C is correct.

The *like dissolves like* rule applies when a solvent is miscible with a solute that has similar properties.

A nonpolar solute is immiscible with a polar solvent.

Retinol (vitamin A)

51. B is correct.

For ionic equations, only aqueous (*aq*) species are broken down into their ions.

Solids, liquids, and gases stay the same.

52. E is correct.

For K_{sp}, only aqueous species are included in the calculation.

The decomposition of $PbCl_2$:

$$PbCl_2 \rightarrow Pb^{2+} + 2\ Cl^-$$

$$K_{sp} = [Pb^{2+}]\cdot[Cl^-]^2$$

When x moles of $PbCl_2$ fully dissociate, x moles of Pb and $2x$ moles of Cl^- are produced:

$$K_{sp} = (x)\cdot(2x)^2$$

$$K_{sp} = 4x^3$$

53. B is correct.

Strong electrolytes dissociate entirely (or almost entirely) in water.

Strong acids and bases dissociate nearly completely (i.e., form stable anions), but weak acids and bases dissociate only slightly (i.e., form unstable anions).

54. A is correct.

A *spectator ion* exists in the same form as reactants and products of a chemical reaction.

The balanced equation for potassium hydroxide and nitric acid:

$$KOH + HNO_3 \rightarrow KNO_3 + H_2O$$

Ionic equation:

$$K^+\,{}^-OH + H^+\,NO_3^- \rightarrow K^+\,NO_3^- + H_2O$$

55. A is correct.

When a solution is diluted, the moles of solute (n) is constant.

However, the molarity and volume change because n = MV:

$$n_1 = n_2$$

$$M_1V_1 = M_2V_2$$

$$V_2 = (M_1V_1) / M_2$$

$$V_2 = (0.20 \text{ M} \times 6.0 \text{ L}) / 14 \text{ M}$$

$$V_2 = 0.086 \text{ L}$$

Convert to milliliters:

$$0.086 \text{ L} \times (1{,}000 \text{ mL} / \text{L}) = 86 \text{ mL}$$

56. B is correct.

When a solution is diluted, the moles of solute are constant.

Use this formula to calculate the new molarity:

$$M_1V_1 = M_2V_2$$

$$160 \text{ mL} \times 4.50 \text{ M} = 595 \text{ mL} \times M_2$$

$$M_2 = (160 \text{ mL} / 595 \text{ mL}) \times 4.50 \text{ M}$$

$$M_2 = 1.21 \text{ M}$$

57. B is correct.

Like dissolves like means that polar substances dissolve in polar solvents and nonpolar substances in nonpolar solvents.

Since benzene is nonpolar, look for a nonpolar substance.

Silver chloride is ionic, while CH_2Cl_2, H_2S and SO_2 are polar.

58. B is correct.

Molarity can vary based on temperature because it involves the volume of the solution.

At different temperatures, the volume of water varies slightly, which affects the molarity.

59. D is correct.

Concentration:

solute / volume

For example: 10 g / 1 liter = 10 g/liter

2 g / 1 liter = 2 g/liter

60. E is correct.

The NaOH content is 5.0% (w/v); the solute (NaOH) is measured in grams, but the solution volume is measured in milliliters.

In this problem, the mass and volume of the solution are interchangeable (1 g = 1 mL).

mass of NaOH = % NaOH × volume of solution

mass of NaOH = 5.0% × 75.0 mL

mass of NaOH = 3.75 g

Practice Set 4: Questions 61–80

61. D is correct.

Adding NaCl increases [Cl$^-$] in solution (i.e., common ion effect), which increases the precipitation of PbCl$_2$ because the ion product increases. This increase causes lead chloride to precipitate and the concentration of free chloride in solution to decrease.

62. A is correct.

AgCl has a stronger tendency to form than PbCl$_2$ because of its smaller K_{sp}.

Therefore, as AgCl forms, an equivalent amount of PbCl$_2$ dissolves.

63. A is correct.

For the dissolution of PbCl$_2$:

$$PbCl_2 \rightarrow Pb^{2+} + 2\ Cl^-$$

$$K_{sp} = [Pb^{2+}]\cdot[2\ Cl^-]^2 = 10^{-5}$$

$$K_{sp} = (x)\cdot(2x)^2 = 10^{-5}$$

$$4x^3 = 10^{-5}$$

$$x \approx 0.014$$

$$[Cl^-] = 2(0.014)$$

$$[Cl^-] = 0.028$$

For the dissolution of AgCl:

$$AgCl \rightarrow Ag^+ + Cl^-$$

$$K_{sp} = [Ag^+]\cdot[Cl^-]$$

$$K_{sp} = (x)\cdot(x)$$

$$10^{-10} = x^2$$

$$x = 10^{-5}$$

$$[Cl^-] = 10^{-5}$$

64. A is correct.

The intermolecular bonding (i.e., van der Waals) in alkanes is similar.

65. C is correct.

Ideally, dilute solutions are so dilute that solute molecules do not interact.

Therefore, the mole fraction of the solvent approaches one.

66. D is correct.

Henry's law states that, at a constant temperature, the amount of gas that dissolves in a volume of liquid is directly proportional to the partial pressure of that gas in equilibrium with that liquid.

Tyndall effect is light scattering by particles in a colloid (or particles in a very fine suspension).

A *colloidal suspension* contains microscopically dispersed insoluble particles (i.e., colloid) suspended throughout the liquid. The colloid particles are larger than those of the solution but not large enough to precipitate due to gravity.

67. C is correct.

Miscible refers to the property (of solutions) when two or more substances (e.g., water and alcohol) are mixed without separating.

68. B is correct.

Start by calculating moles of glucose:

Moles of glucose = mass of glucose / molar mass of glucose

Moles of glucose = 10.0 g / (180.0 g/mol)

Moles of glucose = 0.0555 mol

Then, divide moles by volume to calculate molarity:

Molarity of glucose = moles of glucose / volume of glucose

Molarity of glucose = 0.0555 mol / (100 mL × 0.001 L/mL)

Molarity = 0.555 M

69. E is correct.

An insoluble compound is incapable of being dissolved (especially regarding water).

Hydroxide salts of Group I elements are soluble.

Hydroxide salts of Group II elements (Ca, Sr and Ba) are slightly soluble.

Hydroxide salts of transition metals and Al^{3+} are insoluble.

Thus, $Fe(OH)_3$, $Al(OH)_3$, $Co(OH)_2$ are not soluble.

Most sulfate salts are soluble.

Important exceptions are $BaSO_4$, $PbSO_4$, Ag_2SO_4 and $SrSO_4$.

Salts containing Cl^-, Br^- and I^- are generally soluble.

Exceptions are halide salts of Ag^+, Pb^{2+} and $(Hg_2)^{2+}$. $AgCl$, $PbBr_2$ and Hg_2Cl_2 are insoluble.

Chromates are frequently insoluble. Examples: $PbCrO_4$, $BaCrO_4$

70. C is correct.

The overall reaction is:

$$Zn\ (s) + 2\ HCl\ (aq) \rightarrow ZnCl_2\ (aq) + H_2\ (g)$$

Because HCl and $ZnCl_2$ are ionic, they dissociate into ions in aqueous solutions:

$$Zn\ (s) + 2\ H^+\ (aq) + 2\ Cl^-\ (aq) \rightarrow Zn^{2+}\ (aq) + 2\ Cl^-\ (aq) + H_2\ (g)$$

The common ion (Cl^-) cancels to yield the net ionic reaction:

$$Zn\ (s) + 2\ H^+\ (aq) \rightarrow Zn^{2+}\ (aq) + H_2\ (g)$$

71. E is correct.

The *like dissolves like* rule applies when a solvent is miscible with a solute that has similar properties.

A polar solute (e.g., ketone, alcohols, or carboxylic acids) is miscible with a polar (water) solvent.

The three molecules are polar and are therefore soluble in water.

72. A is correct.

The solubility constant (K_{sp}) calculation is similar to the equilibrium constant.

The concentration of each species is raised to the power of its coefficients, multiplied with each other.

For K_{sp}, only aqueous species are included in the calculation.

$$CaF_2\,(s) \rightarrow Ca^{2+}\,(aq) + 2F^-\,(aq)$$

The concentration of fluorine ions can be determined using the concentration of calcium ions:

$$[F^-] = 2 \times [Ca^{2+}]$$

$$[F^-] = 2 \times 0.00021 \text{ M}$$

$$[F^-] = 4.2 \times 10^{-4} \text{ M}$$

Then, K_{sp} can be determined:

$$K_{sp} = [Ca^{2+}]\cdot[F^-]^2$$

$$K_{sp} = (2.1 \times 10^{-4} \text{ M}) \times (4.2 \times 10^{-4} \text{ M})^2$$

$$K_{sp} = 3.7 \times 10^{-11}$$

73. E is correct.

To determine the number of equivalents:

(4 moles / liter) × (3 equivalents / mole) × (1/3 liter)

= 4 equivalents

74. A is correct.

An electrolyte is a substance that produces an electrically conducting solution when dissolved in a polar solvent (e.g., water).

The dissolved electrolyte separates into positively-charged cations and negatively-charged anions.

Electrolytes conduct electricity.

75. A is correct.

This is a frequently encountered insoluble salt in chemistry problems.

Most sulfate salts are soluble. Important exceptions are $BaSO_4$, $PbSO_4$, Ag_2SO_4 and $SrSO_4$.

76. C is correct.

When the ion product is *equal to or greater* than K_{sp}, precipitation of the salt occurs.

If the ion product value is *less* than K_{sp}, precipitation does *not* occur.

77. C is correct.

Gas solubility is inversely proportional to temperature, so a low temperature increases solubility.

The high pressure of O_2 above the solution increases the pressure of the solution (which improves solubility) and the amount of O_2 available for dissolving.

78. C is correct.

Molality is the number of solute moles dissolved in 1,000 grams of solvent.

The mass of the solution is 1,000 g + mass of solute.

Mass of solute (CH_3OH) = moles CH_3OH × molecular mass of CH_3OH

Mass of solute (CH_3OH) = 8.60 moles × [12.01 g/mol + (4 × 1.01 g/mol) + 16 g/mol]

Mass of solute (CH_3OH) = 8.60 moles × (32.05 g/mol)

Mass of solute (CH_3OH) = 275.63 g

Total mass of solution: 1,000 g + 275.63 g = 1,275.63 g

Volume of solution = mass / density

Volume of solution = 1,275.63 g / 0.94 g/mL

Volume of solution = 1,357.05 mL

Divide moles by volume to calculate molarity:

Molarity = number of moles / volume

Molarity = 8.60 moles / 1,357.05 mL

Molarity = 6.34 M

79. E is correct.

By definition, a 15.0% aqueous solution of KI contains 15% KI and the remainder (100 % – 15% = 85%) is water.

If there are 100 g of KI solution, it has 15 g of KI and 85 g of water.

The answer choice of 15 g KI / 100 g water is incorrect because 100 g is the mass of the solution; the actual mass of water is 85 g.

80. D is correct.

Molarity equals moles of solute divided by liters of solution.

$$1 \text{ cm}^3 = 1 \text{ mL}$$

Molarity:

$$(0.75 \text{ mol}) / (0.075 \text{ L}) = 10 \text{ M}$$

Practice Set 5: Questions 81–100

81. D is correct.

Mass % of solute = (mass of solute / mass of seawater) × 100%

Mass % of solute = (1.35 g / 25.88 g) × 100%

Mass % of solute = 5.22%

82. A is correct.

Some solutions are colored (e.g., Kool-Aid powder dissolved in water).

The *color of chemicals* is a physical property of chemicals from (most commonly) the excitation of electrons due to absorption of energy by the chemical.

The observer sees not the absorbed color but the wavelength that is reflected.

Most simple inorganic (e.g., sodium chloride) and organic compounds (e.g., ethanol) are colorless.

Transition metal compounds are often colored because of transitions of electrons between *d*-orbitals of different energy.

Organic compounds tend to be colored when there is extensive conjugation (i.e., alternating double and single bonds), causing the energy gap between the HOMO (i.e., highest occupied molecular orbital) and LUMO (i.e., lowest unoccupied molecular orbital) to decrease, bringing the absorption band from the UV to the visible region.

Color is due to the energy absorbed by the compound when an electron transitions from the HOMO to the LUMO.

A physical change is a change in physical properties, such as melting, the transition to gas, changes to crystal form, color, volume, shape, size, and density.

83. E is correct.

As the name suggests, the solubility product constant (K_{sp}) is a product of the solubility of each ion in an ionic compound.

Start by writing the dissociation equation of AgCl in a solution:

$$AgCl\ (s) \rightarrow Ag^+\ (aq) + Cl^-\ (aq)$$

The solubility of AgCl is provided in the problem: 1.3×10^{-4} mol/L. This is the maximum amount of AgCl that can be dissolved in water in standard conditions.

There are 1.3×10^{-4} mol AgCl dissolved in one liter of water, which means 1.3×10^{-4} mol of Ag^+ ions and 1.3×10^{-4} mol of Cl^- ions.

K_{sp} is calculated using the same method as the equilibrium constant: ion concentration to the power of ion coefficient. Only aqueous species are included in the calculation.

For AgCl:

$$K_{sp} = [Ag]\cdot[Cl]$$

$$K_{sp} = (1.3 \times 10^{-4})\cdot(1.3 \times 10^{-4})$$

$$K_{sp} = 1.7 \times 10^{-8}$$

84. E is correct.

Concentration:

> solute / volume

For example:

> 10 g / 1 liter = 10 g/liter

> 5 g / 0.5 liter = 10 g/liter

85. B is correct.

The reaction is *endothermic* because the temperature of the solution drops as the reaction absorbs heat from the environment, and $\Delta H°$ is positive.

The solution is unsaturated at 1 molar, so dissolving more salt at standard conditions is spontaneous, and $\Delta G°$ is negative.

86. E is correct.

The *heat of a solution* is the enthalpy change (or energy absorbed as heat at constant pressure) when a solution forms.

For *solution formation*, solvent-solvent bonds and solute-solute bonds must be broken while solute-solvent bonds are formed.

The breaking of bonds absorbs energy, while the formation of bonds releases energy.

If the heat of the solution is negative, energy is released.

87. E is correct.

The reaction is endothermic if the bonds formed have lower energy than the bonds broken.

From $\Delta G = \Delta H - T\Delta S$, the entropy of the system increases if the reaction is spontaneous.

88. B is correct.

Hydrogen bonding in H_2O is stronger than van der Waals's forces in the nonpolar hydrocarbon of benzene.

89. D is correct.

Balanced reaction:

$$BaCl_2 + K_2CrO_4 \rightarrow BaCrO_4 + 2\ KCl$$

Most Cl and K compounds are soluble; therefore, KCl is more likely to be soluble than $BaCrO_4$.

90. C is correct.

Sodium carbonate reacts with calcium and magnesium ions (responsible for the 'hard water' phenomena) and forms precipitates of calcium carbonate and magnesium carbonate. This reduces the hardness of the water.

91. C is correct.

Calculate moles of $NaHCO_3$:

Moles of $NaHCO_3$ = mass $NaHCO_3$ / molar mass $NaHCO_3$

Moles of $NaHCO_3$ = 0.400 g / 84.0 g/mol

Moles of $NaHCO_3$ = 4.76×10^{-3} mol

In this reaction, $NaHCO_3$ and HCl have coefficients of 1.

continued…

Therefore,

$$\text{moles NaHCO}_3 = \text{moles HCl}$$

$$\text{moles NaHCO}_3 = 4.76 \times 10^{-3} \text{ mol.}$$

Use this to calculate volume of HCl:

$$\text{volume of HCl} = \text{moles / molarity}$$

$$\text{volume of HCl} = 4.76 \times 10^{-3} \text{ mol} / 0.25 \text{ M}$$

$$\text{volume of HCl} = 0.019 \text{ L}$$

Convert volume to milliliters:

$$0.019 \text{ L} \times (1{,}000 \text{ mL/L}) = 19.0 \text{ mL}$$

92. D is correct.

Calculation of concentration in ppm (parts per million) is similar to percentage calculation; the difference is that the multiplication factor is 10^6 ppm instead of 100%.

$$\text{Concentration of contaminant} = (\text{mass of contaminant / mass of solution}) \times 10^6 \text{ ppm}$$

$$\text{Concentration of contaminant} = [(8.8 \times 10^{-3} \text{ g}) / (5{,}246 \text{ g})] \times 10^6 \text{ ppm}$$

$$\text{Concentration of contaminant} = 1.68 \text{ ppm}$$

93. A is correct.

The bicarbonate ion is the conjugate base of carbonic acid: H_2CO_3

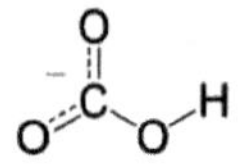

The bicarbonate ion carries a -1 formal charge and is the conjugate base of carbonic acid (H_2CO_3); it is the conjugate acid of the carbonate ion (CO_2^{-3}).

The equilibrium reactions are shown below:

$$CO_3^{2-} + 2\ H_2O \leftrightarrow HCO_3^- + H_2O + OH^- \leftrightarrow H_2CO_3 + 2\ OH^-$$

$$H_2CO_3 + 2\ H_2O \leftrightarrow HCO_3^- + H_3O^+ + H_2O \leftrightarrow CO_3^{2-} + 2\ H_3O^+$$

94. E is correct.

Solutes dissolve in the solvent to form a solution.

95. E is correct.

Hydration involves the interaction of water molecules with the solute.

The water molecules exchange bonding relationships with the solute, whereby water-water bonds break and water-solute bonds form.

96. C is correct.

Hydration involves the interaction of water molecules with the solute.

The water molecules exchange bonding relationships with the solute whereby water-water bonds break and water-solute bonds form.

Hydrates are not new compounds, but rather compound molecules surrounded by water crystals.

A *colloidal suspension* contains microscopically dispersed insoluble particles (i.e., colloid) suspended throughout the liquid.

The colloid particles are larger than those of the solution but not large enough to precipitate due to gravity.

97. A is correct.

Reaction:

$$2 \; AgNO_3 \; (aq) + K_2CrO_4 \; (aq) \rightarrow 2 \; KNO_3 \; (aq) + AgCrO_4 \; (s)$$

Salts containing nitrate ions (NO_3^-) are generally soluble.

98. E is correct.

The *like dissolves like* rule applies when a solvent is miscible with a solute that has similar properties.

A polar solute is miscible with a polar solvent, and the rule applies for the nonpolar solute / solvent.

The three molecules are nonpolar and are therefore miscible with the nonpolar liquid bromine.

The attractive forces between the solutes and the liquid bromine solution are van der Waals.

99. C is correct.

Molecules can have covalent bonds in one situation and ionic bonds in another.

Hydrogen chloride (HCl) is a gas in which hydrogen and chlorine are covalently bound.

However, if HCl is bubbled into the water, it ionizes completely to yield H^+ and Cl^- of a hydrochloric acid solution.

Hydrochloric acid contains two nonmetals and is a covalently bonded molecule.

Hydrochloric acid is a strong acid and dissociates into ions when placed in water (i.e., strong electrolyte).

100. D is correct.

The addition of solute to a liquid lowers the vapor pressure of the liquid because some water molecules are bonding to the solute.

This results in fewer water molecules available for vaporization and therefore reduces the vapor pressure.

Notes for active learning

Notes for active learning

APPENDIX

Periodic Table of the Elements

Key:
atomic number — **Symbol** — name — abridged standard atomic weight

Group	1	2	3	4	5	6	7	8	9	10	11	12	13	14	15	16	17	18
1	1 **H** hydrogen 1.0080 ±0.0002																	2 **He** helium 4.0026 ±0.0001
2	3 **Li** lithium 6.94 ±0.06	4 **Be** beryllium 9.0122 ±0.0001											5 **B** boron 10.81 ±0.02	6 **C** carbon 12.011 ±0.002	7 **N** nitrogen 14.007 ±0.001	8 **O** oxygen 15.999 ±0.001	9 **F** fluorine 18.998 ±0.001	10 **Ne** neon 20.180 ±0.001
3	11 **Na** sodium 22.990 ±0.001	12 **Mg** magnesium 24.305 ±0.002											13 **Al** aluminium 26.982 ±0.001	14 **Si** silicon 28.085 ±0.001	15 **P** phosphorus 30.974 ±0.001	16 **S** sulfur 32.06 ±0.02	17 **Cl** chlorine 35.45 ±0.01	18 **Ar** argon 39.95 ±0.16
4	19 **K** potassium 39.098 ±0.001	20 **Ca** calcium 40.078 ±0.004	21 **Sc** scandium 44.956 ±0.001	22 **Ti** titanium 47.867 ±0.001	23 **V** vanadium 50.942 ±0.001	24 **Cr** chromium 51.996 ±0.001	25 **Mn** manganese 54.938 ±0.001	26 **Fe** iron 55.845 ±0.002	27 **Co** cobalt 58.933 ±0.001	28 **Ni** nickel 58.693 ±0.001	29 **Cu** copper 63.546 ±0.003	30 **Zn** zinc 65.38 ±0.02	31 **Ga** gallium 69.723 ±0.001	32 **Ge** germanium 72.630 ±0.008	33 **As** arsenic 74.922 ±0.001	34 **Se** selenium 78.971 ±0.008	35 **Br** bromine 79.904 ±0.003	36 **Kr** krypton 83.798 ±0.002
5	37 **Rb** rubidium 85.468 ±0.001	38 **Sr** strontium 87.62 ±0.01	39 **Y** yttrium 88.906 ±0.001	40 **Zr** zirconium 91.224 ±0.002	41 **Nb** niobium 92.906 ±0.001	42 **Mo** molybdenum 95.95 ±0.01	43 **Tc** technetium [97]	44 **Ru** ruthenium 101.07 ±0.02	45 **Rh** rhodium 102.91 ±0.01	46 **Pd** palladium 106.42 ±0.01	47 **Ag** silver 107.87 ±0.01	48 **Cd** cadmium 112.41 ±0.01	49 **In** indium 114.82 ±0.01	50 **Sn** tin 118.71 ±0.01	51 **Sb** antimony 121.76 ±0.01	52 **Te** tellurium 127.60 ±0.03	53 **I** iodine 126.90 ±0.01	54 **Xe** xenon 131.29 ±0.01
6	55 **Cs** caesium 132.91 ±0.01	56 **Ba** barium 137.33 ±0.01	57-71 lanthanoids	72 **Hf** hafnium 178.49 ±0.01	73 **Ta** tantalum 180.95 ±0.01	74 **W** tungsten 183.84 ±0.01	75 **Re** rhenium 186.21 ±0.01	76 **Os** osmium 190.23 ±0.03	77 **Ir** iridium 192.22 ±0.01	78 **Pt** platinum 195.08 ±0.02	79 **Au** gold 196.97 ±0.01	80 **Hg** mercury 200.59 ±0.01	81 **Tl** thallium 204.38 ±0.01	82 **Pb** lead 207.2 ±1.1	83 **Bi** bismuth 208.98 ±0.01	84 **Po** polonium [209]	85 **At** astatine [210]	86 **Rn** radon [222]
7	87 **Fr** francium [223]	88 **Ra** radium [226]	89-103 actinoids	104 **Rf** rutherfordium [267]	105 **Db** dubnium [268]	106 **Sg** seaborgium [269]	107 **Bh** bohrium [270]	108 **Hs** hassium [269]	109 **Mt** meitnerium [277]	110 **Ds** darmstadtium [281]	111 **Rg** roentgenium [282]	112 **Cn** copernicium [285]	113 **Nh** nihonium [286]	114 **Fl** flerovium [290]	115 **Mc** moscovium [290]	116 **Lv** livermorium [293]	117 **Ts** tennessine [294]	118 **Og** oganesson [294]

Lanthanoids (57–71)

57 **La** lanthanum 138.91 ±0.01	58 **Ce** cerium 140.12 ±0.01	59 **Pr** praseodymium 140.91 ±0.01	60 **Nd** neodymium 144.24 ±0.01	61 **Pm** promethium [145]	62 **Sm** samarium 150.36 ±0.02	63 **Eu** europium 151.96 ±0.01	64 **Gd** gadolinium 157.25 ±0.03	65 **Tb** terbium 158.93 ±0.01	66 **Dy** dysprosium 162.50 ±0.01	67 **Ho** holmium 164.93 ±0.01	68 **Er** erbium 167.26 ±0.01	69 **Tm** thulium 168.93 ±0.01	70 **Yb** ytterbium 173.05 ±0.02	71 **Lu** lutetium 174.97 ±0.01

Actinoids (89–103)

89 **Ac** actinium [227]	90 **Th** thorium 232.04 ±0.01	91 **Pa** protactinium 231.04 ±0.01	92 **U** uranium 238.03 ±0.01	93 **Np** neptunium [237]	94 **Pu** plutonium [244]	95 **Am** americium [243]	96 **Cm** curium [247]	97 **Bk** berkelium [247]	98 **Cf** californium [251]	99 **Es** einsteinium [252]	100 **Fm** fermium [257]	101 **Md** mendelevium [258]	102 **No** nobelium [259]	103 **Lr** lawrencium [262]

International Union of Pure and Applied Chemistry (IUPAC), 4 May 2022

Notes for active learning

Notes for active learning

Common Chemistry Equations

Throughout the test the following symbols have the definitions specified unless otherwise noted.

L, mL	= liter(s), milliliter(s)		mm Hg	= millimeters of mercury
g	= gram(s)		J, kJ	= joule(s), kilojoule(s)
nm	= nanometer(s)		V	= volt(s)
atm	= atmosphere(s)		mol	= mole(s)

ATOMIC STRUCTURE

$$E = h\nu$$

$$c = \lambda\nu$$

E = energy

ν = frequency

λ = wavelength

Planck's constant, $h = 6.626 \times 10^{-34}$ J s

Speed of light, $c = 2.998 \times 10^8$ m s^{-1}

Avogadro's number $= 6.022 \times 10^{23}$ mol^{-1}

Electron charge, $e = -1.602 \times 10^{-19}$ coulomb

EQUILIBRIUM

$$K_c = \frac{[C]^c[D]^d}{[A]^a[B]^b}, \text{ where } a\,A + b\,B \rightleftarrows c\,C + d\,D$$

$$K_p = \frac{(P_C)^c(P_D)^d}{(P_A)^a(P_B)^b}$$

$$K_a = \frac{[H^+][A^-]}{[HA]}$$

$$K_b = \frac{[OH^-][HB^+]}{[B]}$$

$$K_w = [H^+][OH^-] = 1.0 \times 10^{-14} \text{ at } 25°C$$

$$= K_a \times K_b$$

$$pH = -\log[H^+],\ pOH = -\log[OH^-]$$

$$14 = pH + pOH$$

$$pH = pK_a + \log\frac{[A^-]}{[HA]}$$

$$pK_a = -\log K_a,\ pK_b = -\log K_b$$

Equilibrium Constants

K_c (molar concentrations)

K_p (gas pressures)

K_a (weak acid)

K_b (weak base)

K_w (water)

KINETICS

$$\ln[A]_t - \ln[A]_0 = -kt$$

$$\frac{1}{[A]_t} - \frac{1}{[A]_0} = kt$$

$$t_{1/2} = \frac{0.693}{k}$$

k = rate constant

t = time

$t_{1/2}$ = half-life

GASES, LIQUIDS, AND SOLUTIONS

$$PV = nRT$$

$$P_A = P_{\text{total}} \times X_A, \text{ where } X_A = \frac{\text{moles A}}{\text{total moles}}$$

$$P_{\text{total}} = P_A + P_B + P_C + \ldots$$

$$n = \frac{m}{M}$$

$$K = {}^\circ C + 273$$

$$D = \frac{m}{V}$$

$$KE \text{ per molecule} = \frac{1}{2}mv^2$$

Molarity, M = moles of solute per liter of solution

$$A = abc$$

P = pressure
V = volume
T = temperature
n = number of moles
m = mass
M = molar mass
D = density
KE = kinetic energy
v = velocity
A = absorbance
a = molar absorptivity
b = path length
c = concentration

Gas constant, $R = 8.314 \text{ J mol}^{-1} \text{ K}^{-1}$
$= 0.08206 \text{ L atm mol}^{-1} \text{ K}^{-1}$
$= 62.36 \text{ L torr mol}^{-1} \text{ K}^{-1}$
1 atm $= 760 \text{ mm Hg}$
$= 760 \text{ torr}$
STP $= 0.00\,^\circ C$ and 10^5 Pa

THERMOCHEMISTRY/ ELECTROCHEMISTRY

$$q = mc\Delta T$$

$$\Delta S^\circ = \sum S^\circ \text{ products} - \sum S^\circ \text{ reactants}$$

$$\Delta H^\circ = \sum \Delta H_f^\circ \text{ products} - \sum \Delta H_f^\circ \text{ reactants}$$

$$\Delta G^\circ = \sum \Delta G_f^\circ \text{ products} - \sum \Delta G_f^\circ \text{ reactants}$$

$$\Delta G^\circ = \Delta H^\circ - T\Delta S^\circ$$

$$= -RT \ln K$$

$$= -n F E^\circ$$

$$I = \frac{q}{t}$$

q = heat
m = mass
c = specific heat capacity
T = temperature
S° = standard entropy
H° = standard enthalpy
G° = standard free energy
n = number of moles
E° = standard reduction potential
I = current (amperes)
q = charge (coulombs)
t = time (seconds)

Faraday's constant, $F = 96,485$ coulombs per mole of electrons

$$1 \text{ volt} = \frac{1 \text{ joule}}{1 \text{ coulomb}}$$

Annotated Glossary of Chemistry Terms

A

Absolute entropy (of a substance) – the increase in the entropy of a substance as it goes from a perfectly ordered crystalline form at 0 K (where its entropy is zero) to the temperature in question.

Absolute zero – the zero point on the absolute temperature scale; –273.15 °C or 0 K; theoretically, the temperature at which molecular motion ceases (i.e., the system does not emit or absorb energy, atoms at rest).

Absorption spectrum – spectrum associated with the absorption of electromagnetic radiation by atoms (or other species), resulting from transitions from lower to higher energy states.

Accuracy – the degree to which a value is close to the actual value; see *precision.*

Acid – a substance that produces H^+ (aq) ions in an aqueous solution and gives a pH of less than 7.0; strong acids ionize entirely or almost entirely in dilute aqueous solution; weak acids ionize only slightly. It turns litmus red.

Acid dissociation constant – an equilibrium constant for dissociating a weak acid.

Acid rain – rainwater with a pH of less than 5.7; caused by the gases NO_2 (vehicle exhaust fumes) and SO_2 (from burning fossil fuels) dissolving in the rain. It kills fish, wildlife, and trees and destroys buildings and lakes.

Acidic salt – contains an ionizable hydrogen atom; does not necessarily produce acidic solutions.

Actinides – the fifteen chemical elements that are between actinium (89) and lawrencium (103).

Activated complex – a structure formed by a collision between molecules in which new bonds form.

Activation energy – the amount of energy that reactants must absorb in their ground states to reach the transition state needed for a reaction to occur.

Active metal – a metal with low ionization energy that loses electrons readily to form cations.

Activity (of a component of ideal mixture) – a dimensionless quantity whose magnitude is equal to the molar concentration in an ideal solution; equal to partial pressure in an ideal gas mixture; 1 for pure solids or liquids.

Activity series – a listing of metals (and hydrogen) in order of decreasing activity.

Actual yield – the amount of a specified pure product obtained from a given reaction; see *theoretical yield.*

Addition reaction – a reaction in which two atoms or groups of atoms are added to a molecule, one on each side of a double or triple bond.

Adhesive forces – forces of attraction between a liquid and another surface.

Adsorption – the adhesion of a species onto the surfaces of particles.

Aeration – the mixing of air into a liquid or a solid.

Alcohol – a hydrocarbon derivative containing a ~OH group attached to a carbon atom, not in an aromatic ring.

Alkali metals – the elements of Group IA on the periodic table (e.g., Na, K, Rb).

Alkaline battery – a dry cell in which the electrolyte contains KOH.

Alkaline earth metals – group IIA metals on the periodic table; see *earth metals*.

Allomer – a substance that has a different composition from another but the same crystalline structure.

Allotropes – elements with different structures (therefore different forms), such as carbon (e.g., diamond, graphite, and fullerene).

Allotropic modifications (allotropes) – different forms of the same element in the same physical state.

Alloy – a mixture of metals. For example, bronze is an alloy formed from copper and tin.

Alloying – mixing metals with other substances (usually other metals) to modify their properties.

Alpha (α) particle – a helium nucleus; helium ion with 2+ charge; an assembly of two protons and two neutrons.

Amorphous solid – a non-crystalline solid with no well-defined ordered structure.

Ampere – unit of electrical current; one ampere equals one coulomb per second.

Amphiprotism – the ability of a substance to exhibit amphiprotic by accepting donated protons.

Amphoterism – the ability to react with both acids and bases; to act as either an acid or a base.

Amplitude – the maximum distance that medium particles carrying the wave move from their rest position.

Anion – a negative ion; an atom or group of atoms that has gained one or more electrons.

Anode – in a cathode ray tube, the positive electrode (electrode at which oxidation occurs); the positive side of a dry cell battery or a cell.

Antibonding orbital – a molecular orbital higher in energy than any of the atomic orbitals from which it is derived; lends instability to a molecule or ion when populated with electrons; denoted with star (*) superscript.

Artificial transmutation – an artificially induced nuclear reaction caused by the bombardment of a nucleus with subatomic particles or small nuclei.

Associated ions – short-lived species formed by the collision of dissolved ions of opposite charges.

Atmosphere – a unit of pressure; the pressure supports a column of mercury 760 mm high at 0 °C.

Atom – a chemical element in its smallest form; it comprises neutrons and protons within the nucleus and electrons circling the nucleus; it is the smallest part of an element that can exist.

Atomic mass unit (amu) – one-twelfth of the mass of an atom of the carbon-12 isotope; used for stating atomic and formula weights; known as a dalton.

Atomic number – represents an element corresponding with the number of protons within the nucleus; the number of protons in the nucleus of the atom.

Atomic orbital (*AO*) – a region or volume in space where the probability of finding electrons is highest.

Atomic radius – radius of an atom.

Atomic weight – weighted average of the masses of the constituent isotopes of an element; the relative masses of atoms of different elements.

Aufbau (or *building up*) principle – describes the order in which electrons fill orbitals in atoms.

Autoionization – an ionization reaction between identical molecules.

Avogadro's Law – equal volumes of gases contain the same number of molecules at the same temperature and pressure.

Avogadro's number (N_A) – the number (6.022×10^{23}) of atoms, molecules, or particles found in precisely 1 mole of a substance.

B

Background radiation – extraneous to an experiment; usually the low-level natural radiation from cosmic rays and trace radioactive substances present in the environment.

Band – a series of very closely spaced nearly continuous molecular orbitals that belong to the crystal as a whole.

Band of stability – band containing nonradioactive nuclides in a plot of neutrons *vs.* their atomic number.

Band theory of metals – the theory that accounts for the bonding and properties of metallic solids.

Barometer – a device used to measure the pressure in the atmosphere.

Base – a substance that produces $^-$OH (*aq*) ions in an aqueous solution; accepts a proton and has a high pH; strongly soluble bases are soluble in water and are entirely dissociated; weak bases ionize only slightly; a typical example of a base is sodium hydroxide (NaOH). It turns litmus blue.

Basic anhydride – the oxide of a metal that reacts with water to form a base.

Basic salt – a salt containing an ionizable OH group.

Beta (β) particle – an electron emitted from the nucleus when a neutron decays to a proton and an electron.

Binary acid – a binary compound in which H is bonded to one or more electronegative nonmetals.

Binary compound – consists of two elements; it may be ionic or covalent.

Binding energy (nuclear binding energy) – the energy equivalent ($E = mc^2$) of the mass deficiency of an atom (where E is the energy in joules, m is the mass in kilograms, and c is the speed of light in m/s^2).

Boiling – the phase transition of a liquid vaporizing.

Boiling point – the temperature at which the vapor pressure of a liquid is equal to the applied pressure; the *condensation point*.

Boiling point elevation – the increase in the boiling point of a solvent caused by the dissolution of a nonvolatile solute.

Bomb calorimeter – a device used to measure the heat transfer between a system and its surroundings at constant volume.

Bond – the attraction and repulsion between atoms and molecules is a cornerstone of chemistry.

Bond energy – the amount of energy necessary to break one mole of bonds in a substance, dissociating the substance in its gaseous state into atoms of its elements in the gaseous state.

Bond order – half the number of electrons in bonding orbitals minus half the electrons in antibonding orbitals.

Bonding orbital – a molecular orbit lower in energy than any of the atomic orbitals from which it is derived; lends stability to a molecule or ion when populated with electrons.

Bonding pair – pair of electrons involved in a covalent bond.

Boron hydrides – binary compounds of boron and hydrogen.

Born-Haber cycle – a series of reactions (and the accompanying enthalpy changes) which, when summed, represent the hypothetical one-step reaction by which elements in their standard states are converted into crystals of ionic compounds (and the accompanying enthalpy changes).

Boyle's Law – at a constant temperature, the volume occupied by a definite mass of a gas is inversely proportional to the applied pressure.

Breeder reactor – a nuclear reactor that produces more fissionable nuclear fuel than it consumes.

Brønsted-Lowrey acid – a chemical species that donates a proton.

Brønsted-Lowrey base – a chemical species that accepts a proton.

Buffer solution – resists change in pH; contains either a weak acid and a soluble ionic salt of the acid or a weak base and a soluble ionic salt of the base.

Buret – a piece of volumetric glassware, usually graduated in 0.1 mL intervals, used to deliver solutions for titrations in a quantitative (drop-like) manner; also spelled *burette*.

C

Calorie – the amount of heat required to raise the temperature of one gram of water from 14.5 °C to 15.5 °C; 1 calorie = 4.184 joules.

Calorimeter – a device used to measure the heat transfer between a system and its surroundings.

Canal ray – a stream of positively charged particles (cations) that moves toward the negative electrode in cathode ray tubes; observed to pass through canals in the negative electrode.

Capillary – a tube having a very small inside diameter.

Capillary action – the drawing of a liquid up the inside of a small-bore tube when adhesive forces exceed cohesive forces; the depression of the surface of the liquid when cohesive forces exceed the adhesive forces.

Catalyst – a chemical compound used to change the rate (to speed or slow it) of a regenerated reaction (i.e., not consumed) at the end of the reaction.

Catenation – the bonding of atoms of the same element into chains or rings (i.e., the ability of an element to bond with itself).

Cathode – the electrode at which reduction occurs; in a cathode ray tube, the negative electrode.

Cathodic protection – protection of a metal (making a cathode) against corrosion by attaching it to a sacrificial anode of a more easily oxidized metal.

Cathode ray tube – a closed glass tube containing gas under low pressure, with electrodes near the ends and a luminescent screen near the positive electrode; produces cathode rays when a high voltage is applied.

Cation – a positive ion; an atom or group of atoms that has lost one or more electrons.

Cell potential – the potential difference, E_{cell}, between oxidation and reduction half-cells under nonstandard conditions; the force in a galvanic cell pulls electrons through a reducing agent to an oxidizing agent.

Central atom – an atom in a molecule or polyatomic ion bonded to more than one other atom.

Chain reaction – a reaction that, once initiated, sustains itself and expands; a reaction in which reactive species, such as radicals, are produced in more than one step, allowing these reactive species to propagate the chain reaction.

Charles' Law – at constant pressure, the volume occupied by a definite mass of gas is directly proportional to its absolute temperature.

Chemical bonds – the attractive forces holding atoms together in elements or compounds.

Chemical change – when one or more new substances are formed.

Chemical equation – description of a chemical reaction by placing the formulas of the reactants on the left of an arrow and the formulas of the products on the right.

Chemical equilibrium – a state of dynamic balance in which the rates of forward and reverse reactions are equal; there is no net change in concentrations of reactants or products while a system is at equilibrium.

Chemical kinetics – studies rates and mechanisms of chemical reactions and factors they depend on.

Chemical periodicity – the variations in properties of elements with their position in the periodic table.

Chemical reaction – the change of one or more substances into another or multiple substances.

Cloud chamber – a device for observing the paths of speeding particles as vapor molecules condense on them to form fog-like tracks.

Cobalt chloride paper – water test; water changes the color from blue to pink.

Coefficient of expansion – the ratio of the change in the length or the volume of a body to the original length or volume for a unit change in temperature.

Cohesive forces – the forces of attraction among particles of a liquid.

Colligative properties – physical properties of solutions that depend upon the number but not the kind of solute particles present.

Collision theory – the theory of reaction rates that states that effective collisions between reactant molecules must occur for the reaction to occur.

Colloid – a heterogeneous mixture in which solute-like particles do not settle (e.g., many kinds of milk).

Combination reaction – two substances (elements or compounds) combine to form one compound.

Combustible – classification of liquid substances that burn based on flashpoints; any liquid having a flashpoint at or above 37.8 °C (100 °F) but below 93.3 °C (200 °F), except any mixture having components with flashpoints of 93.3 °C (200 °F) or higher, the total of which makes up 99% or more of the volume of the mixture.

Combustion (or *burning*) – an exothermic reaction between an oxidant and fuel with heat and often light.

Common ion effect – suppression of ionization of a weak electrolyte by the presence in the same solution of a strong electrolyte containing one of the same ions as the weak electrolyte.

Complex ions – ions resulting from coordinating covalent bonds between simple ions and other ions or molecules.

Composition stoichiometry – describes the quantitative (mass) relationships among elements in compounds.

Compound – a substance of two or more chemically bonded elements in fixed proportions; can be decomposed into constituent elements.

Compressed gas – a single or mixture of gases having (in a container) an absolute pressure exceeding 40 psi at 21.1 °C (70 °F).

Compression – an area in a longitudinal wave where the particles are closer and pushed in.

Concentration – the amount of solute per unit volume, the mass of solvent or solution.

Condensation – the phase change from gas to liquid.

Condensed phases – the liquid and solid phases; phases in which particles interact strongly.

Condensed states – the solid and liquid states.

Conduction – heat transfer between substances in direct contact with each other (i.e., must be touching); when particles of a hotter substance vibrate, these molecules bump into nearby particles and transfer some energy.

Conduction band – a vacant or partially filled band of energy levels just higher in energy than a filled band; a band within which, or into which, electrons must be promoted to allow electrical conduction to occur in a solid.

Conductor – a material that allows electric flow more freely.

Conjugate acid-base pair – in Brønsted-Lowry terms, reactant and product that differ by a proton, H^+.

Conformations – structures of a compound that differ by the extent of rotation about a single bond.

Continuous spectrum – contains wavelengths in a specified region of the electromagnetic spectrum.

Control rods – rods of materials such as cadmium or boron steel that act as neutron absorbers (not merely moderators), used in nuclear reactors to control neutron fluxes and therefore fission rates.

Conjugated double bonds – double bonds separated from each other by one single bond –C=C–C=C–

Contact process – the industrial process for sulfur trioxide (SO_3) and sulfuric acid (H_2SO_4) production from sulfur dioxide (SO_2).

Convection – the physical flow of matter when heat flows by energized molecules from one place to another through the movement of fluids. Transfer of heat through a liquid or a gas occurs when molecules of the liquid or gas move and carry the heat.

Coordinate covalent bond – a covalent bond with shared electrons furnished by the same species; a bond between a Lewis acid and a Lewis base.

Coordination compound or complex – a compound containing coordinate covalent bonds.

Coordination number – the number of donor atoms coordinated to a metal; the number of nearest neighbors of an atom or ion in describing crystals.

Coordination sphere – the metal ion and its coordinating ligands, but no uncoordinated counter-ions.

Corrosion – oxidation of metals (e.g., rusting) in the presence of air and moisture.

Coulomb – the SI unit of electrical charge; unit symbol – C.

Covalent bond – a force of attraction (chemical bond) formed by the sharing of electron pairs between two atoms.

Covalent compounds – compounds made of two or more nonmetal atoms bonded by sharing valence electrons.

Critical mass – the minimum mass of a particular fissionable nuclide in a given volume required to sustain a nuclear chain reaction.

Critical point – the combination of critical temperature and critical pressure of a substance.

Critical pressure – the pressure required to liquefy a gas (vapor) at its *Critical temperature.*

Critical temperature – the temperature above which a gas cannot be liquefied; the temperature above which a substance cannot exhibit distinct gas and liquid phases.

Crystal – a solid packed with ions, molecules, or atoms in an orderly fashion.

Crystal field stabilization energy – a measure of the net energy of stabilization gained by a metal ion's nonbonding d electrons due to complex formation.

Crystal field theory – bonding in transition metal complexes in which ligands and metal ions are treated as point charges; a purely ionic model; ligand point charges represent the crystal (electrical) field perturbing the metal's d orbitals containing nonbonding electrons.

Crystal lattice – a pattern of arrangement of particles in a crystal.

Crystal lattice energy – the amount of energy that holds a crystal together; the energy change when a mole of solid forms from its constituent molecules or ions (for ionic compounds) in their gaseous state (consistently negative).

Crystalline solid – a solid characterized by a regular, ordered arrangement of particles.

Curie (Ci) – the basic unit to describe the intensity of radioactivity in a sample of material; one curie equals 37 billion disintegrations per second or approximately the amount of radioactivity given off by 1 gram of radium.

Current – a flow of charged particles, such as electrons or ions, moving through an electrical space or conductor. It is measured as the net rate of flow of electric charge; the unit is an Ampere (A).

Cuvette – glassware used in spectroscopic experiments; usually made of plastic, glass, or quartz, and should be as clean and transparent as possible.

Cyclotron – a device for accelerating charged particles along a spiral path.

D

Dalton's Law (or the *law of partial pressures*) – the pressure exerted by a mixture of gases is the sum of the partial pressures of the individual gases.

Daughter nuclide – nuclide produced in nuclear decay.

Debye – the unit used to express dipole moments.

Degenerate – in orbitals, describes orbitals of the same energy.

Deionization – the removal of ions; in the case of water, mineral ions such as sodium, iron, and calcium.

Deliquescence – substances that absorb water from the atmosphere to form liquid solutions.

Delocalization – in reference to electrons, bonding electrons distributed among more than two atoms bonded; occurs in species that exhibit resonance.

Density – mass per unit volume; $D = M \times V$.

Deposition – settling particles within a solution; the direct solidification of vapor by cooling; see *sublimation*.

Derivative – a compound that can be imagined arising from a parent compound by replacing one atom with another atom or group of atoms; used extensively in organic chemistry to identify compounds.

Detergent – a soap-like emulsifier with sulfate, SO_3, or a phosphate group instead of carboxylate group.

Deuterium – an isotope of hydrogen whose atoms are twice as massive as ordinary hydrogen; deuterium atoms contain a proton and a neutron in the nucleus.

Dextrorotatory – refers to an optically active substance that rotates plane-polarized light clockwise, also known as *dextro*.

Diagonal similarities – chemical similarities in the Periodic Table of Elements of Period 2 to elements of Period 3 one group to the right, especially evident toward the left of the periodic table.

Diamagnetism – weak repulsion by a magnetic field.

Differential Scanning Calorimetry (DSC) – a technique for measuring temperature, direction, and magnitude of thermal transitions in a sample material by heating/cooling and comparing the amount of energy required to maintain its rate of temperature increase or decrease with an inert reference material under similar conditions.

Differential Thermal Analysis (DTA) – a technique for observing the temperature, direction, and magnitude of thermally induced transitions in a material by heating/cooling a sample and comparing its temperature with an inert reference material under similar conditions.

Differential thermometer – a thermometer used to measure minimal temperature changes accurately.

Dilution – the process of reducing the concentration of a solute in a solution, usually by mixing with more solvent.

Dimer – a molecule formed by combining two smaller (identical) molecules.

Dipole – electric or magnetic separation of charge; charge separation between two covalently bonded atoms.

Dipole-dipole interactions – attractive electrostatic forces between polar molecules (i.e., between molecules with permanent dipoles).

Dipole moment – the product of the distance separating opposite charges of equal magnitude; a measure of the polarity of a bond or molecule; a measured dipole refers to the dipole moment of an entire molecule.

Dispersing medium – the solvent-like phase in a colloid.

Dispersed phase – the solute-like species in a colloid.

Displacement reactions – reactions in which one element displaces another from a compound.

Disproportionation reactions – redox reactions in which the oxidizing agent and the reducing agent are the same species.

Dissociation – in an aqueous solution, the process by which a solid ionic compound separates into its ions.

Dissociation constant – equilibrium constant for dissociating a complex ion into a simple ion and coordinating species (ligands).

Dissolution or solvation – the spread of ions in a monosaccharide.

Distilland – the material in a distillation apparatus that is to be distilled.

Distillate – the material in a distillation apparatus collected in the receiver.

Distillation – separating a liquid mixture into its components based on differences in boiling points; the process in which components of a mixture are separated by boiling away the more volatile liquid; the vaporization of a liquid by heating and then the condensation of the vapor by cooling.

Domain – a cluster of atoms in a ferromagnetic substance, which align in the same direction in the presence of an external magnetic field.

Donor atom – a ligand atom whose electrons are shared with a Lewis acid.

***d*-orbitals** – beginning in the third energy level, a set of five degenerate orbitals per energy level, higher in energy than *s* and *p* orbitals of the same energy level.

Dosimeter – a small, calibrated electroscope worn by laboratory personnel to measure incident ionizing radiation or chemical exposure.

Double bond – covalent bond resulting from the sharing of four electrons (two pairs) between two atoms.

Double salt – a solid consisting of two co-crystallized salts.

Doublet – two peaks or bands of about equal intensity appearing close on a spectrogram.

Downs cell – an electrolytic cell for the commercial electrolysis of molten sodium chloride.

DP number – the degree of polymerization; the average number of monomer units per polymer unit.

Dry cells (voltaic cells) – ordinary batteries for appliances (e.g., flashlights, radios).

Dumas method – a method used to determine the molecular weights of volatile liquids.

Dynamic equilibrium – an equilibrium in which the processes occur continuously with no net change.

E

Earth metal – highly reactive elements in group IIA of the periodic table (includes beryllium, magnesium, calcium, strontium, barium, and radium); see *alkaline earth metal*.

Effective collisions – a collision between molecules resulting in a reaction, one in which the molecules collide with proper relative orientations and sufficient energy to react.

Effective molality – the sum of the molalities of solute particles in a solution.

Effective nuclear charge – the nuclear charge experienced by the outermost electrons of an atom; the actual nuclear charge minus the effects of shielding due to inner-shell electrons (e.g., a set of $dx_2\text{-}y_2$ and dz_2 orbitals); those d orbitals within a set with lobes directed along the x, y, and z axes.

Electrical conductivity – the measure of how easily an electric current can flow through a substance.

Electric charge – a measured property (coulombs) that determines electromagnetic interaction.

Electrochemical cell – using a chemical reaction, an electromotive force is generated.

Electrochemistry – the study of chemical changes produced by electrical current and electricity production by chemical reactions.

Electrodes – surfaces upon which oxidation and reduction half-reactions occur in electrochemical cells; a conductor dips into an electrolyte and allows the electrons to flow to and from the electrolyte.

Electrode potentials – potentials, E, of half-reactions as reductions vs. standard hydrogen electrode.

Electrolysis – 1) occurs in electrolytic cells; chemical decomposition occurs by passing an electric current through a solution containing ions. 2) producing a chemical change using electricity; used to split up water into H and O_2.

Electrolyte – a substance (i.e., anions, cations) which, when dissolved in water, conducts electricity. An ionic solution that conducts a certain amount of current is categorized into two types: weak and strong.

Electrolytic cells – electrochemical cells in which electrical energy causes nonspontaneous redox reactions to occur (i.e., forced to occur by applying an outside source of electrical energy).

Electrolytic conduction – electrical current passes through ions in a solution or pure liquid.

Electromagnetic radiation – energy propagated using electric and magnetic fields that oscillate in directions perpendicular to the direction of travel of the energy; a type of wave that can go through vacuums as well as material; classified as a "self-propagating wave."

Electromagnetism – fields of an electric charge and electric properties that change how particles move and interact.

Electromotive force – a device that gains energy as electric charges pass through it.

Electromotive series – the relative order of tendencies for elements and their simple ions to act as oxidizing or reducing agents; also known as the "activity series."

Electron – a subatomic particle having a mass of 0.00054858 amu and a charge of –1.

Electron affinity – the energy absorbed in the process in which an electron is added to a neutral isolated gaseous atom to form a gaseous ion with a 1– charge; it has a negative value if energy is released.

Electron configuration – the specific distribution of electrons in atomic orbitals of atoms or ions.

Electron-deficient compounds – at least one atom (other than H) shares fewer than eight electrons.

Electron shells – an orbital around an atom's nucleus with a fixed number of electrons (usually 2 or 8).

Electronic transition – the transfer of an electron from one energy level to another.

Electronegativity – a measure of the relative tendency of an atom to attract electrons to itself when chemically combined with another atom.

Electronic geometry – the geometric arrangement of orbitals containing the shared and unshared electron pairs surrounding the central atom or polyatomic ion.

Electrophile – positively charged or electron-deficient.

Electrophoresis – separating ions by migration rate and direction of migration in an electric field.

Electroplating – a metal is covered with another metal layer using electricity; plating a metal onto a (cathodic) surface by electrolysis.

Element – a substance that cannot be decomposed into simpler substances by chemical means; defined by its *atomic number*. A substance that cannot be split into simpler substances by chemical means.

Eluant (or eluent) – the solvent used in the process of elution, as in liquid chromatography.

Eluate – a solvent (or mobile phase) which passes through a chromatographic column and removes the sample components from the stationary phase.

Emission spectrum – the emission of electromagnetic radiation by atoms (or other species) resulting from electronic transitions from higher to lower energy states.

Empirical formula – the simplest whole-number ratio of atoms of each element present in a compound; also known as the *simplest formula*.

Emulsifying agent – a substance that coats the particles of the dispersed phase and prevents coagulation of colloidal particles; an emulsifier.

Emulsion – colloidal suspension of a liquid in a liquid.

Endothermic – describes processes that absorb heat energy (*H*).

Endothermicity – the absorption of heat by a system as the process occurs.

Endpoint – the point at which an indicator changes color and titration stops.

Energy – a system's ability to do work.

Enthalpy (*H*) – the heat content of a specific amount of substance; E= PV.

Entropy *(S)* – a thermodynamic state or property that measures the degree of disorder (i.e., randomness) of a system; the amount of energy not available for work in a closed thermodynamic system (usually denoted by S).

Enzyme – a protein that acts as a catalyst in biological systems.

Equation of state – an expression that describes the behavior of matter in a given state; the van der Waals equation describes the behavior of the gaseous state.

Equilibrium or chemical equilibrium – a state of dynamic balance with the rates of forward and reverse reactions equal; the state of a system where neither forward nor reverse reaction is thermodynamically favored.

Equilibrium constant – a quantity characterizing equilibrium position for a reversible reaction; its magnitude is equal to mass action expression at equilibrium; equilibrium, "K," varies with temperature.

Equivalence point – the point when chemically equivalent amounts of reactants have reacted.

Equivalent weight – an oxidizing or reducing agent whose mass gains (oxidizing agents) or loses (reducing agents) 6.022×10^{23} electrons in a redox reaction.

Evaporation – vaporization of a liquid below its boiling point.

Evaporation rate – the rate at which a particular substance vaporizes (evaporate) compared to the rate of a known substance, such as ethyl ether, especially useful for health and fire-hazard considerations.

Excited state – any state other than the ground state of an atom or molecule; see *ground state*.

Exothermic – describes processes that release heat energy (*H*).

Exothermicity – the release of heat by a system as a process occurs.

Explosive – a chemical or compound that causes a sudden, almost instantaneous release of pressure, gas, heat, and light when subjected to sudden shock, pressure, high temperature, or applied potential.

Explosive limits – the range of concentrations over which a flammable vapor mixed with the proper ratios of air will ignite or explode if a source of ignition is provided.

Extensive property – a property that depends upon the amount of material in a sample.

Extrapolate – to estimate the value of a result outside the range of a series of known values; a technique used in the standard additions calibration procedure.

F

Faraday constant – a unit of electrical charge widely used in electrochemistry and equal to ~ 96,500 coulombs; represents 1 mole of electrons, or the Avogadro number of electrons: 6.022×10^{23} electrons.

Faraday's law of electrolysis – a two-part law that Michael Faraday published about electrolysis. 1. the mass of a substance altered at an electrode during electrolysis is directly proportional to the quantity of electricity transferred at that electrode. 2. the mass of an elemental material altered at an electrode is directly proportional to the element's equivalent weight; one equivalent weight of a substance is produced at each electrode during the passage of 96,487 coulombs of charge through an electrolytic cell.

Fast neutron – a neutron ejected at high kinetic energy in a nuclear reaction.

Ferromagnetism – the ability of a substance to become permanently magnetized by exposure to an external magnetic field.

Flashpoint – the temperature at which a liquid yields enough flammable vapor to ignite; various recognized industrial testing methods exist; therefore, the method used must be specified.

Fluorescence – absorption of high-energy radiation and subsequent emission of visible light.

First Law of Thermodynamics – the amount of energy in the universe is constant (i.e., energy is neither created nor destroyed in ordinary chemical reactions and physical changes); known as the Law of Conservation of Energy.

Fluids – substances that flow freely; gases and liquids.

Flux – a substance added to react with the charge or a product of its reduction; in metallurgy, it is usually added to lower the melting point.

Foam – colloidal suspension of a gas in a liquid.

Formal charge – a method of counting electrons in a covalently bonded molecule or ion; it counts bonding electrons as though they were equally shared between the two atoms.

Formula – a combination of symbols that indicates the chemical composition of a substance.

Formula unit – the smallest repeating unit of a substance; the molecule for nonionic substances.

Formula weight – the mass of one formula unit of a substance in atomic mass units.

Fossil fuels – formed from the remains of plants and animals that lived millions of years ago.

Fractional distillation – when a fractioning column is used in a distillation apparatus to separate the components of a liquid mixture with different boiling points.

Fractional precipitation – removal of some ions from a solution by precipitation while leaving other ions with similar properties in the solution.

Free energy change – the indicator of the spontaneity of a process at constant T and P (e.g., if ΔG is negative, the process is spontaneous).

Free radical – a highly reactive chemical species carrying no charge and having a single unpaired electron in an orbital.

Freezing – phase transition from liquid to solid.

Freezing point depression – the decrease in the freezing point of a solvent caused by the presence of a solute.

Frequency – the number of repeating points on a wave that passes a given observation point per unit time; the unit is 1 hertz = 1 cycle per 1 second.

Fuel – any substance that burns in oxygen to produce heat.

Fuel cells – a voltaic cell that converts the chemical energy of a fuel and an oxidizing agent directly into electrical energy continuously.

G

Gamma (γ) ray – a highly penetrating type of nuclear radiation similar to X-ray radiation, except that it comes from within the nucleus of an atom and has higher energy; energy-wise, very similar to cosmic rays, except that cosmic rays originate from outer space.

Galvanic cell – a battery made from an electrochemical cell with two metals connected by a salt bridge.

Galvanizing – placing a thin layer of zinc on a ferrous material to protect the underlying surface from corrosion.

Gangue – sand, rock, and other impurities surrounding the mineral of interest in an ore.

Gas – a state of matter in which the particles have no definite shape or volume, though they fill their container.

Gay-Lussac's Law – the expression used for each of the two relationships named after the French chemist Joseph Louis Gay-Lussac concerning the properties of gases; more usually applied to his law of combining volumes.

Geiger counter – a gas-filled tube that discharges electrically when ionizing radiation passes through it.

Gel – colloidal suspension of a solid dispersed in a liquid; a semi-rigid solid.

Gibbs (free) energy – the thermodynamic state function of a system that indicates the amount of energy available for the system to do useful work at constant T and P; a value that indicates the spontaneity of a reaction (usually denoted by G).

Graham's Law – the rates of effusion of gases are inversely proportional to the square roots of their molecular weights or densities.

Ground state – the lowest energy state or most stable state of an atom, molecule, or ion; see *excited state*.

Group – a vertical column in the periodic table; known as a family.

H

Haber process – a process for the catalyzed industrial production of ammonia from N_2 and H_2 at high temperature and pressure.

Half-cell – the compartment in which the oxidation or reduction half-reaction occurs in a voltaic cell.

Half-life – the time required for half of a reactant to be converted into product(s); the time required for half of a given sample to undergo radioactive decay.

Half-reaction – the oxidation or the reduction part of a redox reaction.

Halogens – group VIIA elements: F, Cl, Br, I; halogens are nonmetals.

Hard water – water high in dissolved minerals that makes it difficult to form lather with soap.

Heat – a form of energy that flows between two samples of matter because of temperature differences.

Heat capacity – the amount of heat required to raise the temperature of a mass one degree Celsius.

Heat of condensation – the amount of heat that must be removed from one gram of vapor at its condensation point to condense the vapor with no change in temperature.

Heat of crystallization – the amount of heat that must be removed from one gram of a liquid at its freezing point to freeze it with no change in temperature.

Heat of fusion – the amount of heat required to melt one gram of a solid at its melting point with no change in temperature; usually expressed in J/g; the molar heat of fusion is the amount of heat required to melt one mole of a solid at its melting point with no change in temperature and is usually expressed in kJ/mol.

Heat of solution – the amount of heat absorbed in forming a solution that contains one mole of the solute; the value is positive if heat is absorbed (endothermic) and negative if heat is released (exothermic).

Heat of vaporization – the amount of heat required to vaporize one gram of a liquid at its boiling point with no change in temperature; usually expressed in J/g; the molar heat of vaporization is the amount of heat required to vaporize one mole of liquid at its boiling point with no change in temperature and is usually expressed as ion kJ/mol.

Heisenberg uncertainty principle – states that it is impossible to determine the momentum and position of an electron simultaneously with absolute accuracy.

Henry's Law – the gas pressure above a solution is proportional to concentration of the gas in solution.

Hess' Law of heat summation – the enthalpy change for a reaction is the same whether it occurs in one step or a series of steps.

Heterogeneous catalyst – exists in a different phase (solid, liquid, or gas) from the reactants; a contact catalyst.

Heterogeneous equilibria – equilibria involving species in more than one phase.

Heterogeneous mixture – a mixture that does not have uniform composition and properties throughout.

Heteronuclear – consisting of different elements.

High spin complex – crystal field designation for an outer orbital complex; t_{2g} and e_g orbitals are singly occupied before pairing occurs.

Homogeneous catalyst – in the same phase (solid, liquid, or gas) as the reactants.

Homogeneous equilibria – when *reagents* and *products* are in the same phase (gases, liquids, or solids).

Homogeneous mixture – a mixture which has uniform composition and properties throughout.

Homologous series – compounds with each member differing from the next by a specific number and kind of atoms.

Homonuclear – consisting of only one element.

Hund's rule – single electrons must occupy orbitals of a given sublevel before pairing begins; see *Aufbau* (or *building up*) *principle*.

Hybridization – mixing atomic orbitals to form a new set of atomic orbitals with the same electron capacity and properties and energies intermediate between the original unhybridized orbitals.

Hydrate – a solid compound that contains a definite percentage of bound water.

Hydrate isomers – crystalline complexes that differ in whether water exists inside or outside the coordination sphere.

Hydration – the reaction of a substance with water.

Hydration energy – the energy change accompanying the hydration of a mole of gas and ions.

Hydride – a binary compound of hydrogen.

Hydrocarbons – compounds that contain only carbon and hydrogen.

Hydrogen bond – a relatively strong dipole-dipole interaction (but still considerably weaker than the covalent or ionic bonds) between molecules containing hydrogen directly bonded to a small, highly electronegative atom, such as N, O or F.

Hydrogenation – the reaction in which hydrogen adds across a double or triple bond.

Hydrogen-oxygen fuel cell – hydrogen is the fuel (reducing agent), and oxygen is the oxidizing agent.

Hydrolysis – the reaction of a substance with water or its ions.

Hydrolysis constant – an equilibrium constant for a hydrolysis reaction.

Hydrometer – a device used to measure the densities of liquids and solutions.

Hydrophilic colloids – colloidal particles that repel water molecules.

I

Ideal gas – a hypothetical gas that obeys the postulates of the Kinetic Molecular Theory.

Ideal gas law – the product of pressure and the volume of an ideal gas is directly proportional to the number of moles of the gas and the absolute temperature. $PV = nRT$

Ideal solution – obeys Raoult's Law strictly.

Immiscible liquids – do not mix to form a solution (e.g., oil and water).

Indicators – for acid-base titrations, organic compounds that exhibit different colors in solutions of different acidities, determine the point at which the reaction between two solutes is complete.

Inert pair effect – characteristic of the post-transition minerals; the tendency of the electrons in the outermost atomic *s* orbital to remain un-ionized or unshared in compounds of post-transition metals.

Inhibitory catalyst – an inhibitor; a catalyst that decreases the rate of reaction.

Inner orbital complex – valence bond designation for a complex in which the metal ion utilizes d orbitals for one shell inside the outermost occupied shell in its hybridization.

Inorganic chemistry – a part of chemistry concerned with inorganic (non-carbon-based) compounds.

Insulator – a material that resists the flow of electric current or heat transfer; it does not allow heat to flow easily.

Insoluble compound – a substance that will not dissolve in a solvent, even after mixing.

Integrated rate equation – an expression giving the concentration of a reactant remaining after a specified time; it has a different mathematical form for different orders of reactants.

Intermolecular forces – forces between individual particles (atoms, molecules, ions) of a substance.

Ion – a molecule that has gained or lost electrons; an atom or a group of atoms carries an electric charge (Na^+).

Ion product for water – equilibrium constant for water ionization; $Kw = [H_3O^+]\cdot[^-OH] = 1.00 \times 10^{-14}$ at 25 °C.

Ionic bond – the electrostatic attraction between oppositely charged ions, resulting from a transfer of electrons.

Ionic bonding – chemical bonding resulting from transferring electrons from one atom or group.

Ionic compounds – compounds containing predominantly ionic bonding.

Ionic geometry – arrangement of atoms (not lone pairs of electrons) about the central atom of a polyatomic ion.

Ionization – the breaking up of a compound into separate ions; in an aqueous solution, the process by which a molecular compound reacts with water and forms ions.

Ionization constant – equilibrium constant for the ionization of a weak electrolyte.

Ionization energy – the minimum amount of energy required to remove the most loosely held electron of an isolated gaseous atom or ion.

Ion exchange – a method of removing hardness from water, it replaces the positive ions that cause the hardness with H^+ ions.

Ionization isomers – result from the interchange of ions inside and outside the coordination sphere.

Isoelectric – having the same electronic configurations.

Isomers – different substances with the same formula.

Isomorphous – refers to crystals having the same atomic arrangement.

Isotopes – two or more forms of atoms of the same element with different masses; atoms containing the same number of protons but different numbers of neutrons.

IUPAC – acronym for "International Union of Pure and Applied Chemistry."

J

Joule (J) – a unit of energy in the SI system; one joule is $1 \; kg\cdot m^2/s^2$, which is 0.2390 calories.

K

K capture – absorption of a K shell (n = 1) electron by a proton as it is converted to a neutron.

Kelvin – a unit of measure for temperature based upon an absolute scale.

Kinetics – a subfield of chemistry specializing in reaction rates.

Kinetic energy (*KE*) – energy that matter processes through its motion.

Kinetic Molecular Theory – a theory that attempts to explain macroscopic observations on gases in microscopic or molecular terms.

L

Lanthanides – elements 57 (lanthanum) through 71 (lutetium); grouped because of their similar behavior in chemical reactions.

Lanthanide contraction – a decrease in the radii of the elements following the lanthanides compared to what would be expected if there were no f-transition metals.

Latent heat – the energy absorbed or released when a substance changes state without changing temperature.

Lattice – a unique arrangement of atoms or molecules in a crystalline liquid or solid.

Law of combining volumes (Gay-Lussac's Law) – at constant temperature and pressure, the volumes of reacting gases (and any gaseous products) can be expressed as ratios of small whole numbers.

Law of conservation of energy – energy cannot be created or destroyed; it can only change form.

Law of conservation of matter – there is no detectable change in the quantity of matter during an ordinary chemical reaction.

Law of conservation of matter and energy – the amount of matter and energy in the universe is fixed.

Law of definite proportions (law of constant composition) – different samples of a pure compound contain the same elements in the same proportions by mass.

Law of partial pressures (or *Dalton's Law*) – the pressure exerted by a mixture of gases is the sum of the partial pressures of the individual gases.

Laws of thermodynamics – physical laws which define quantities of thermodynamic systems describe how they behave and (by extension) set certain limitations, such as perpetual motion.

Lead storage battery – a secondary voltaic cell used in most automobiles.

Leclanche cell – a common type of *dry cell*.

Le Chatelier's principle – states that a system at equilibrium, or striving to attain equilibrium, responds in such a way as to counteract any stress placed upon it; if stress (change of conditions) is applied to a system at equilibrium, the system will shift in the direction that reduces stress.

Leveling effect – acids stronger than the acid characteristic of the solvent react with the solvent to produce that acid; a similar statement applies to bases. The strongest acid (base) that can exist in a given solvent is the acid (base) characteristic of the solvent.

Levorotatory (or *levo*) – an optically active substance rotates plane-polarized light counterclockwise.

Lewis acid – any species that can accept a share in an electron pair.

Lewis base – any species that can make available a pair of electrons.

Lewis dot formula (electron dot formula) – representation of a molecule, ion, or formula unit by showing atomic symbols and only outer shell electrons.

Ligand – a Lewis base in a coordination compound.

Light – that portion of the electromagnetic spectrum visible to the naked eye; known as "visible light."

Limiting reactant – a substance that stoichiometrically limits the number of products formed.

Linear accelerator – a device used for accelerating charged particles along a straight line path.

Line spectrum – an atomic emission or absorption spectrum.

Linkage isomers – a particular ligand bonds to a metal ion through different donor atoms.

Liquid – a state of matter which takes the shape of its container.

Liquid aerosol – colloidal suspension of a liquid in a gas.

London dispersion forces – very weak and very short-range attractive forces between short-lived temporary (induced) dipoles; known as "dispersion forces."

Lone pair – a pair of electrons residing on one atom and not shared by other atoms; an unshared pair.

Low spin complex – crystal field designation for an inner orbital complex; contains electrons paired t_{2g} orbitals before e_g orbitals are occupied in octahedral complexes.

Lubricant – a substance capable of reducing friction (i.e., a force that opposes the direction of motion).

M

Magnetic field – a space around a magnet where magnetism can be detected.

Magnetic quantum number – quantum mechanical solution to a wave equation designating the orbital within a given set (s, p, d, f) in which an electron resides.

Manometer – a two-armed barometer.

Mass (m) – a measure of the amount of matter in an object; mass is usually measured in grams or kilograms.

Mass action expression – for a reversible reaction, aA + bB cC + dD; the product of the concentrations of the products (species on the right), each raised to the power corresponding to its coefficient in the balanced chemical equation, divided by the product of the concentrations of reactants (species on the left), each raised to the power corresponding to its coefficient in the balanced equation; at equilibrium the mass action expression is equal to K.

Mass deficiency – the amount of matter converted into energy when an atom forms from constituent particles.

Mass number – the sum of the numbers of protons and neutrons in an atom; an integer.

Mass spectrometer – an instrument that measures the charge-to-mass ratio of charged particles.

Matter – anything that has mass and occupies space.

Mechanism – the sequence of steps by which reactants are converted into products.

Melting point – the temperature at which liquid and solid coexist in equilibrium.

Meniscus – the shape assumed by the surface of a liquid in a cylindrical container.

Melting – the phase change from a solid to a liquid.

Metal – a chemical element that is a good conductor of electricity and heat and forms cations and ionic bonds with nonmetals; elements below and to the left of the stepwise division (metalloids) in the upper right corner of the periodic table; about 80% of known elements are metals.

Metallic bonding – bonding within metals due to the electrical attraction of positively charged metal ions for mobile electrons that belong to the crystal.

Metallic conduction – conduction of electrical current through a metal or along a metallic surface.

Metalloid – a substance with the properties of metals and nonmetals (B, Al, Si, Ge, As, Sb, Te, Po and At).

Metathesis reactions – reactions in which two compounds react to form two new compounds, with no changes in oxidation number; reactions in which the ions of two compounds exchange partners.

Method of initial rates – method of determining the rate-law expression by carrying out a reaction with different initial concentrations and analyzing the resultant changes in initial rates.

Methylene blue – a heterocyclic aromatic chemical compound with the molecular formula $C_{16}H_{18}N_3SCl$.

Miscible liquids – mix to form a solution (e.g., alcohol and water).

Miscibility – the ability of one liquid to mix with (dissolve in) another liquid.

Mixture – two or more different substances mingled together but not chemically combined. A sample of matter composed of two or more substances, each of which retains its identity and properties.

Moderator – a substance (e.g., deuterium, oxygen, paraffin) that slows fast neutrons upon collision.

Molality – a concentration expressed as the number of moles of solute per kilogram of solvent.

Molarity – the number of moles of solute per liter of solution.

Molar solubility – the number of moles of a solute that dissolves to produce a liter of a saturated solution.

Mole – a measurement of an amount of substance; a single mole contains approximately 6.022×10^{23} units or entities; abbreviated mol.

Molecule – a chemically bonded number of electrically neutral atoms.

Molecular equation – a chemical reaction in which formulas are written as if substances existed as molecules; only complete formulas are used.

Molecular formula – indicates the number of atoms present in a molecule of a molecular substance.

Molecular geometry – the arrangement of atoms (not lone pairs of electrons) around a central atom of a molecule or polyatomic ion.

Molecular orbital (*MO*) – resulting from the overlap and mixing of atomic orbitals on different atoms (i.e., a region where an electron can be found in a molecule, as opposed to an atom); an MO belongs to the molecule.

Molecular orbital theory – a theory of chemical bonding based upon postulated molecular orbitals.

Molecular weight – the mass of one molecule of a nonionic substance in atomic mass units.

Molecule – the smallest particle of a compound capable of stable, independent existence.

Mole fraction – the number of moles of a component in a mixture divided by the number of moles in the mixture.

Monoprotic acid – can form only one hydronium ion per molecule; may be strong or weak.

Mother nuclide – nuclide that undergoes nuclear decay.

N

Native state – refers to the occurrence of an element in an uncombined or free state in nature.

Natural radioactivity – spontaneous decomposition of an atom.

Neat – conditions with a liquid reagent or gas performed with no added solvent or co-solvent.

Nernst equation – corrects standard electrode potentials for nonstandard conditions.

Net ionic equation – results from canceling spectator ions and eliminating brackets from a total ionic equation.

Neutralization – the reaction of an acid with a base to form a salt and water; usually, the reaction of hydrogen ions with hydroxide ions to form water molecules.

Neutrino – a particle that travels at speeds close to the speed of light; created due to radioactive decay.

Neutron – a neutral unit or subatomic particle with no net charge and a mass of 1.0087 amu.

Nickel-cadmium cell (NiCd battery) – a dry cell where the anode is Cd, the cathode is NiO_2, and the electrolyte is basic.

Nitrogen cycle – the complex series of reactions by which nitrogen is slowly but continually recycled in the atmosphere, lithosphere, and hydrosphere.

Noble gases – elements of the periodic Group 0; He, Ne, Ar, Kr, Xe, Rn; known as "rare gases;" formerly called "inert gases."

Nodal plane – a region in which the probability of finding an electron is zero.

Nonbonding orbital – a molecular orbital derived only from an atomic orbital of one atom; it lends neither stability nor instability to a molecule or ion when populated with electrons.

Nonelectrolyte – a substance whose aqueous solutions do not conduct electricity.

Nonmetal – an element that is not metallic.

Nonpolar bond – a covalent bond in which electron density is symmetrically distributed.

Nuclear – of or about the atomic nucleus.

Nuclear binding energy – the energy equivalent of the mass deficiency; the energy released in forming an atom from the subatomic particles.

Nuclear fission – when a heavy nucleus splits into nuclei of intermediate masses, and protons are emitted.

Nuclear magnetic resonance spectroscopy – a technique that exploits the magnetic properties of specific nuclei; helps identify unknown compounds.

Nuclear reaction – involves a change in the composition of a nucleus and can emit or absorb a tremendous amount of energy.

Nuclear reactor – a system in which controlled nuclear fission reactions generate heat energy on a large scale, subsequently converted into electrical energy.

Nucleons – particles comprising the nucleus: protons and neutrons.

Nucleus – the tiny and dense, positively charged center of an atom containing protons and neutrons, as well as other subatomic particles; the net charge is positive.

Nuclides – refers to different atomic forms of elements; in contrast to isotopes, which refer only to different atomic forms of a single element.

Nuclide symbol – designation for an atom A/Z E, in which E is the symbol of an element, Z is its atomic number, and A is its mass number.

Number density – a measure of the concentration of countable objects (e.g., atoms, molecules) in space; the number per volume.

O

Octahedral – molecules and polyatomic ions with one atom in the center and six atoms at the corners of an octahedron.

Octane number – a number that indicates how smoothly a gasoline burns.

Octet rule – during bonding, atoms tend to reach an electron arrangement with eight electrons in the outermost shell. Many representative elements attain at least a share of eight electrons in their valence shells when they form molecular or ionic compounds; there are some limitations.

Open sextet – species with only six electrons in the highest energy level of the central element (many Lewis acids).

Orbital – may refer to an atomic orbital or a molecular orbital.

Organic chemistry – the chemistry of substances that contain carbon-hydrogen bonds.

Organic compound – substances that contain carbon.

Osmosis – when solvent molecules pass through a semi-permeable membrane from a dilute solution into a more concentrated solution.

Osmotic pressure – the hydrostatic pressure produced on the surface of a semi-permeable membrane.

Outer orbital complex – valence bond designation for a complex in which the metal ion utilizes d orbitals in the outermost (occupied) shell in hybridization.

Overlap – the interaction of orbitals on different atoms in the same region of space.

Oxidation – the addition of oxygen or the loss of electrons. An algebraic increase in the oxidation number may correspond to a loss of electrons.

Oxidation numbers – quantitative values used as mechanical aids in writing formulas and balancing equations; for single-atom ions, they correspond to the charge on the ion; more electronegative atoms are assigned negative oxidation numbers, known as *oxidation states*.

Oxidation-reduction reactions – reactions in which oxidation and reduction occur; known as *redox reactions*.

Oxide – a binary compound of oxygen.

Oxidizing agent – the substance that oxidizes another substance and is reduced.

P

Pairing – a favorable interaction of two electrons with opposite m values in the same orbital.

Pairing energy – the energy required to pair two electrons in the same orbital.

Paramagnetism – attraction toward a magnetic field, stronger than diamagnetism but still weak compared to ferromagnetism.

Partial pressure – the force exerted by one gas in a mixture of gases.

Particulate matter – fine, divided solid particles suspended in polluted air.

Pauli exclusion principle – no electrons in the same atom may have identical sets of four quantum numbers.

Percentage ionization – the percentage of the weak electrolyte that will ionize in a solution of given concentration.

Percent by mass – 100% times the actual yield divided by the theoretical yield.

Percent composition – the mass percent of each element in a compound.

Percent purity – the percent of a specified compound or element in an impure sample.

Period – the elements in a horizontal row of the periodic table.

Periodicity – regular periodic variations of properties of elements with their atomic number (and position in the periodic table).

Periodic Law – the properties of the elements are periodic functions of their atomic numbers.

Periodic table – an arrangement of elements by increasing atomic numbers, emphasizing periodicity.

Peroxide – a compound with oxygen in –1 oxidation state; metal peroxides contain the peroxide ion, O_2^{2-}.

pH – the measure of acidity (or basicity) of a solution; negative logarithm of the concentration (mol/L) of the H_3O^+ [H^+] ion; scale is commonly used over a range 0 to 14.

Phase diagram – shows the equilibrium temperature-pressure relationships for different phases of a substance.

pH scale – a range from 0 to 14. If the pH of a solution is 7, it is neutral; if the pH of a solution is less than 7, it is acidic; if the pH of a solution is greater than 7, it is basic.

Permanent hardness – hardness (relative to lathering soap) in water that cannot be removed by boiling; caused by calcium sulfate.

Photoelectric effect – emission of an electron from the surface of a metal caused by impinging electromagnetic radiation of specific minimum energy; the current increases with increasing radiation intensity.

Photon – a carrier of electromagnetic radiation of all wavelengths, such as gamma rays and radio waves; known as a *quantum of light*.

Physical change – when a substance changes from one physical state to another, but no substances with different compositions are formed; physical change may involve a phase change (e.g., melting, freezing) or another physical change, such as crushing a crystal or separating one volume of liquid into different containers; it does not produce a new substance.

Plasma – a physical state of matter that exists at extremely high temperatures in which molecules are dissociated, and most atoms are ionized.

Polar bond – a covalent bond with an unsymmetrical distribution of electron density.

Polarimeter – a device used to measure optical activity.

Polarization – the buildup of a product of oxidation or reduction of an electrode, preventing further reaction.

Polydentate – refers to ligands with more than one donor atom.

Polyene – a compound that contains more than one double bond per molecule.

Polymerization – the combination of many small molecules to form large molecules.

Polymer – a large molecule consisting of chains or rings of linked monomer units, usually characterized by high melting and boiling points.

Polymorphous – refers to substances that can crystallize in more than one crystalline arrangement.

Polyprotic acid – forms two or more H_3O_+ ions per molecule; often, at least one ionization step is weak.

Positron – a nuclear particle with the mass of an electron but opposite charge (positive).

Potential difference (or *voltage*) – the force that moves the electrons around circuit; unit is Volt (V).

Potential energy (*PE*) – energy stored in a body or a system due to its position in a force field or configuration.

Power – the rate at which energy is converted from one form to another; the unit is Watts (W). Power = voltage × current (P = VI).

Precipitate – an insoluble solid formed by mixing in solution the constituent ions of a slightly soluble solution.

Precision – how close the results of multiple experimental trials are; see *accuracy*.

Pressure – force per unit area; unit is Pascal (Pa).

Primary standard – a known high degree of purity substance that undergoes one invariable reaction with the other reactant of interest.

Primary voltaic cells – voltaic cells that cannot be recharged; no further chemical reaction is possible once the reactants are consumed.

Products – chemicals produced (from reactants) in a chemical reaction.

Proton – a subatomic particle having a mass of 1.0073 amu and a charge of +1, found atom's nucleus.

Protonation – the addition of a proton (H^+) to an atom, molecule, or ion.

Pseudobinaryionic compounds – contain more than two elements but are named like binary compounds.

Q

Quanta – the minimum amount of energy emitted by radiation.

Quantum mechanics – the study of how atoms, molecules, subatomic particles, etc., behave and are structured; a mathematical method of treating particles based on quantum theory, which assumes that energy (of small particles) is not infinitely divisible.

Quantum numbers – numbers that describe the energies of electrons in atoms; derived from quantum mechanical treatment.

Quarks – elementary particles and a fundamental constituent of matter, combining to form hadrons (i.e., protons and neutrons).

R

Radiation – 1) heat transfer through invisible rays, which travel outwards from the hot object without a medium. 2) high-energy particles or rays emitted during the nuclear decay processes.

Radical – an atom or group of atoms that contains one or more unpaired electrons; usually a very reactive species.

Radioactive dating – a method of dating ancient objects by determining the ratio of mother and daughter nuclides present in an object and relating the ratio to the object's age via half-life calculations.

Radioactive tracer – a small amount of radioisotope replacing a nonradioactive isotope of the element in a compound whose path (e.g., in the body) or whose decomposition products are monitored by detection of radioactivity; known as a "radioactive label."

Radioactivity – the spontaneous disintegration of atomic nuclei.

Raoult's Law – the vapor pressure of a solvent in an ideal solution decreases as its mole fraction decreases.

Rate-determining step – the slowest step in a mechanism; determines the overall reaction rate.

Rate-law expression – an equation relating the reaction rate to the concentrations of the reactants and the specific rate of the reaction.

Rate of reaction – the change in the concentration of a reactant or product per unit time.

Reactants – substances consumed in a chemical reaction; react together in a chemical reaction.

Reaction quotient – the mass action expression under any set of conditions (not necessarily equilibrium); its magnitude relative to K determines the direction in which the reaction must occur to establish equilibrium.

Reaction ratio – the relative amounts of reactants and products involved in a reaction; may be the ratio of moles, millimoles, or masses.

Reaction stoichiometry – describes the quantitative relationships among substances participating in chemical reactions.

Reactivity series (or activity series) – an empirical, calculated, and structurally analytical progression of a series of metals, arranged by "reactivity" from highest to lowest; used to summarize information about the reactions of metals with acids and water, double displacement reactions, and the extraction of metals from ores.

Reagent – a substance (or compound) added to a system to cause a chemical reaction or to visualize if a reaction occurs; the terms reactant and reagent are often used interchangeably; however, a *reactant* is more specifically a substance consumed during a chemical reaction.

Reducing agent – a substance that reduces another substance and is itself oxidized.

Reduction – the removal of oxygen or the gaining of electrons.

Resonance – the concept in which two or more equivalent dot formulas for the same arrangement of atoms (resonance structures) are necessary to describe the bonding in a molecule or ion.

Reverse osmosis – forcing solvent molecules to flow through a semi-permeable membrane from a concentrated solution into a dilute solution by applying greater hydrostatic pressure on the concentrated side than the osmotic pressure opposing it.

Reversible reaction – processes that do not go to completion and occur in the forward and reverse direction.

S

Saline solution – a general term for NaCl (i.e., sodium chloride) in water.

Salt – when a metal replaces the hydrogen of an acid.

Salts – ionic compounds composed of anions and cations.

Salt bridge – a U-shaped tube containing an electrolyte, connects the two half-cells of a voltaic cell.

Saturated solution –no more solute will dissolve at that temperature.

s-block elements – group 1 and 2 elements (alkali and alkaline metals), including hydrogen and helium.

Schrödinger equation – quantum state equation representing the behavior of an electron around an atom; describes the wave function of a physical system evolving.

Second Law of Thermodynamics – the universe tends toward a state of greater disorder in spontaneous processes.

Secondary standard – a solution that has been titrated against a primary standard; a standard solution.

Secondary voltaic cells – voltaic cells that can be recharged; original reactants can be regenerated by reversing the direction of the current flow.

Semiconductor – a substance that does not conduct electricity at low temperatures but will do so at higher temperatures.

Semi-permeable membrane – a thin partition between two solutions through which specific molecules can pass but others cannot.

Shielding effect – electrons in filled sets of s, p orbitals between the nucleus and outer shell electrons shield the outer shell electrons somewhat from the effect of protons in the nucleus; known as the "screening effect."

Sigma (σ) bonds – bonds resulting from the head-on overlap of atomic orbitals. The region of electron sharing is along and (cylindrically) symmetrical to the imaginary line connecting the bonded atoms.

Sigma orbital – molecular orbital resulting from the head-on overlap of two atomic orbitals.

Single bond – covalent bond resulting from the sharing of two electrons (one pair) between two atoms.

Sol – a suspension of solid particles in a liquid; artificial examples include sol-gels.

Solid – one of the states of matter, where the molecules are packed closely, resistance to movement/deformation, and volume change.

Solubility product constant – equilibrium constant for the dissolution of a slightly soluble compound.

Solubility product principle – the solubility product constant expression for a slightly soluble compound is the product of the concentrations of the constituent ions; each raised to the power that corresponds to the number of ions in one formula unit.

Solute – the dispersed (i.e., dissolved) phase of a solution; the solution is mixed into the solvent (e.g., NaCl in saline water).

Solution – a homogeneous mixture of multiple substances; comprised of solutes and solvents; a mixture of a solute (usually a solid) and a solvent (usually a liquid).

Solvation – the process by which solvent molecules surround and interact with solute ions or molecules.

Solvent – the dispersing medium of a solution (e.g., H_2O in saline water).

Solvolysis – the reaction of a substance with the solvent in which it is dissolved.

s-**orbital** – a spherically symmetrical atomic orbital; one per energy level.

Specific gravity – the ratio of the density of a substance to the density of water.

Specific heat – the amount of heat required to raise the temperature of one gram of substance 1 °C.

Specific rate constant – an experimentally determined (proportionality) constant; different for different reactions, and which changes only with temperature; k in the rate-law expression: Rate $= k$ [A] $\times$ [B].

Spectator ions – ions in a solution that do not participate in a chemical reaction.

Spectral line – any of several lines corresponding to definite wavelengths of an atomic emission or absorption spectrum; marks the energy difference between two energy levels.

Spectrochemical series – arrangement of ligands in order of increasing ligand field strength.

Spectroscopy – the study of radiation and matter, such as X-ray absorption and emission spectroscopy.

Spectrum – display of component wavelengths (colors) of electromagnetic radiation.

Speed of light – the speed at which radiation travels through a vacuum (299,792,458 m/sec).

Square planar – describes molecules and polyatomic ions with one atom in the center and four atoms at the corners of a square.

Square planar complex – relationship with metal in the center of a square plane, with ligand donor atoms at each of the four corners.

Standard conditions for temperature and pressure (STP) – used to compare experimental results (25 °C and 100.000 kPa).

Standard electrodes – half-cells in which the oxidized and reduced forms of a species are present at the unit activity (1.0 M solutions of dissolved ions, 1.0 atm partial pressure of gases, pure solids, and liquids).

Standard electrode potential – by convention, the potential (Eo) of a half-reaction as a reduction relative to the standard hydrogen electrode when species are present at unit activity.

Standard entropy – the absolute entropy of a substance in its standard state at 298 K.

Standard molar enthalpy of formation – the amount of heat absorbed in forming one mole of a substance in a specified state from its elements in their standard states.

Standard molar volume – space occupied by 1 mole of ideal gas under standard conditions; 22.4 liters.

Standard reaction – a process where the numbers of moles of reactants in the balanced equation, in their standard states, are entirely converted to the numbers of moles of products in the balanced equation, at their standard state.

State of matter – a homogeneous, macroscopic phase (e.g., gas, plasma, liquid, solid) in increasing concentration.

Stoichiometry – quantitative relationships of elements and compounds undergoing chemical changes.

Strong electrolyte – a substance that conducts electricity well in a dilute aqueous solution.

Strong field ligand – a ligand that exerts a strong crystal or ligand electrical field and generally forms low-spin complexes with metal ions when possible.

Structural isomers – compounds that contain the same number and kinds of atoms, but with different geometries.

Subatomic particles – comprise an atom (e.g., protons, neutrons, electrons).

Sublimation – the direct vaporization of a solid by heating without passing through the liquid state; a phase transition from solid to gas.

Substance – any matter, specimens with the same chemical composition and physical properties.

Substitution reaction – a reaction in which another atom or group of atoms replaces an atom or a group of atoms.

Supercooled liquids – liquids that, when cooled, apparently solidify but continue to flow very slowly under the influence of gravity.

Supercritical fluid – a substance at a temperature above its critical temperature.

Supersaturated solution – contains a higher than saturation concentration of solute; slight disturbance or seeding causes crystallization of excess solute.

Suspension – a heterogeneous mixture in which solute-like particles settle out of the solvent-like phase sometime after their introduction. A mixture of a liquid and a finely divided insoluble solid.

T

Talc – a mineral representing the Mohs Scale, composed of hydrated magnesium silicate with the chemical formula $H_2Mg_3(SiO_3)_4$ or $Mg_3Si_4O_{10}(OH)_2$.

Temperature – a measure of heat intensity (i.e., hotness or coldness of a sample); a measure of kinetic energy of an object. Units are measured in degrees, and scales include Celsius, Fahrenheit, and Kelvin.

Temporary hardness – hardness in water, removed by boiling; caused by calcium hydrogen carbonate.

Ternary acid – a ternary compound containing H, O and another element, often a nonmetal.

Ternary compound – a compound consisting of three elements; may be ionic or covalent.

Tetrahedral – a term used to describe molecules and polyatomic ions with one atom in the center and four atoms at the corners of a tetrahedron.

Theoretical yield – the maximum amount of a specified product that could be obtained from specified amounts of reactants, assuming complete consumption of the limiting reactant according to only one reaction and complete recovery of the product; see *actual yield*.

Theory – a model describing the nature of a phenomenon.

Thermal conductivity – a property of a material to conduct heat (often noted as k).

Thermal cracking – decomposition by heating a substance in the presence of a catalyst and without air.

Thermochemistry – the study of absorption/release of heat within a chemical reaction; studies heat energy associated with chemical reactions and physical transformations.

Thermodynamics – studying the effects of changing temperature, volume, or pressure (or work, heat, and energy) on a macroscopic scale.

Thermodynamic stability – when a system is in its lowest energy state with its environment (equilibrium).

Thermometer – a device that measures the average energy of a system.

Thermonuclear energy – energy from nuclear fusion reactions.

Third Law of Thermodynamics – entropy of a pure crystalline substance at absolute zero equals zero.

Titration – a procedure in which one solution is added to another solution until the chemical reaction between the two solutions is complete; the concentration of one solution is known, and that of the other is unknown. The process of adding one solution to a measured amount of another to find out exactly how much of each is required to react.

Torr – a unit to measure pressure; 1 Torr is equivalent to 133.322 Pa or 1.3158×10^{-3} atm.

Total ionic equation – the expression for a chemical reaction written to show the predominant form of species in aqueous solution or contact with water.

Transition elements (metals) – B Group elements except IIB in the periodic table; sometimes called transition elements, elements with incomplete d sub-shells; the d-block elements.

Transition state theory –reactants pass through high-energy transition states before forming products.

Transuranic element – an atomic number greater than 92; none of the transuranic elements are stable.

Triple bond – the sharing of three pairs of electrons within a covalent bond (e.g., N_2).

Triple point – where the temperature and pressure of three phases are the same; water has a unique phase diagram.

Tyndall effect – results from light scattering by colloidal particles (a mixture where one substance is dispersed evenly throughout another) or by suspended particles.

U

Uncertainty –any measurement that involves estimating any amount that cannot be precisely reproduced.

Uncertainty principle – knowing the location of a particle makes the momentum uncertain, while knowing the momentum of a particle makes the location uncertain.

Unit cell – the smallest repeating unit of a lattice.

Unit factor – statements used in converting between units.

Universal (or ideal) gas constant – proportionality constant in the ideal gas law (0.08206 L·atm/(K·mol)).

UN number – a four-digit code used to note hazardous and flammable substances.

Unsaturated hydrocarbons – hydrocarbons that contain double or triple carbon-carbon bonds.

V

Valence bond theory – proposes that covalent bonds are formed when atomic orbitals on different atoms overlap and the electrons are shared.

Valence electrons – outermost electrons of atoms; usually those involved in bonding.

Valence shell electron pair repulsion theory (VSEPR) – assumes electron pairs are arranged around the central element of a molecule or polyatomic ion with maximum separation (and minimum repulsion) among regions of high electron density.

Valency – the number of electrons an atom wants to gain, lose, or share to have a full outer shell.

Van der Waals' equation – a quantitative relationship of a state extending the ideal gas law to real gases by including two empirically determined parameters, which are specific for different gases.

Van der Waals force – one of the forces (attraction/repulsion) between molecules.

Van't Hoff factor – the ratio of moles of particles in solution to moles of solute dissolved.

Vapor – when a substance is below the critical temperature in the gas phase.

Vaporization – the phase change from liquid to gas.

Vapor pressure – the particle pressure of vapor at the surface of its parent liquid.

Viscosity – the resistance of a liquid to flow (e.g., oil has a higher viscosity than water).

Volt – one joule of work per coulomb; the unit of electrical potential transferred.

Voltage – the potential difference between two electrodes; a measure of the chemical potential for a redox reaction.

Voltaic cells – electrochemical cells in which spontaneous chemical reactions produce electricity; known as *galvanic cells*.

Voltmeter – an instrument that measures the cell potential.

Volumetric analysis – measuring the volume of a solution (of known concentration) to determine the substance's concentration within the solution; see *titration*.

W

Water equivalent – the amount of water absorbing the same heat as the calorimeter per degree of temperature increase.

Weak electrolyte – a substance that conducts electricity poorly in a dilute aqueous solution.

Weak field ligand – a ligand that exerts a weak crystal or ligand field and generally forms high-spin complexes with metals.

X

X-ray – electromagnetic radiation between gamma and UV rays.

X-ray diffraction – a method for establishing structures of crystalline solids using single-wavelength X-rays and studying the diffraction pattern.

X-ray photoelectron spectroscopy – a spectroscopic technique used to measure the composition of a material.

Y

Yield – the amount of product produced during a chemical reaction.

Z

Zone melting – remove impurities from an element by melting and slowly traveling it down an ingot (cast).

Zone refining – a method of purifying a metal bar by passing it through an induction heater; this causes impurities to move along a melted portion.

Zwitterion (formerly called a dipolar ion) – a neutral molecule with a positive and negative electrical charge; multiple positive and negative charges can be present, distinct from dipoles at different locations within that molecule; known as *inner salts*.

Frank J. Addivinola, Ph.D.

This study guide's lead author and chief editor is Dr. Frank Addivinola. With his outstanding education, laboratory research, and decades of university science teaching, Dr. Addivinola lent his expertise to oversee the development of this series.

During his extensive career, Dr. Addivinola held faculty positions at colleges and universities, including Harvard University, Johns Hopkins University, University of Maryland, and Northeastern University, and taught undergraduate and graduate-level courses in biology, biochemistry, organic chemistry, inorganic chemistry, physics, anatomy and physiology, medical terminology, nutrition, and medical ethics. He received several awards for his research and presentations.

Dr. Frank Addivinola conducted original research in developmental biology as a doctoral candidate and pre-IRTA fellow in Molecular and Cell Biology at the National Institutes of Health (NIH). His dissertation advisor was Nobel laureate Marshall W. Nirenberg, Chief of the Biochemical Genetics Laboratory at the National Heart, Lung, and Blood Institute (NHLBI). Before NIH, Dr. Addivinola researched prostate cancer in the Cell Growth and Regulation Laboratory of Dr. Arthur Pardee at the Dana Farber Cancer Institute of Harvard Medical School.

Dr. Addivinola holds an undergraduate degree in biology from Williams College. He completed his Masters at Harvard University, Masters in Biotechnology at Johns Hopkins University, and five other graduate degrees at the University of Maryland University College, Suffolk University, and Northeastern University.

Essential Chemistry Self-Teaching Guides

Electronic Structure & Periodic Table

Chemical Bonding

States of Matter & Phase Equilibria

Stoichiometry

Solution Chemistry

Chemical Kinetics & Equilibrium

Acids & Bases

Chemical Thermodynamics

Electrochemistry

Visit our Amazon store

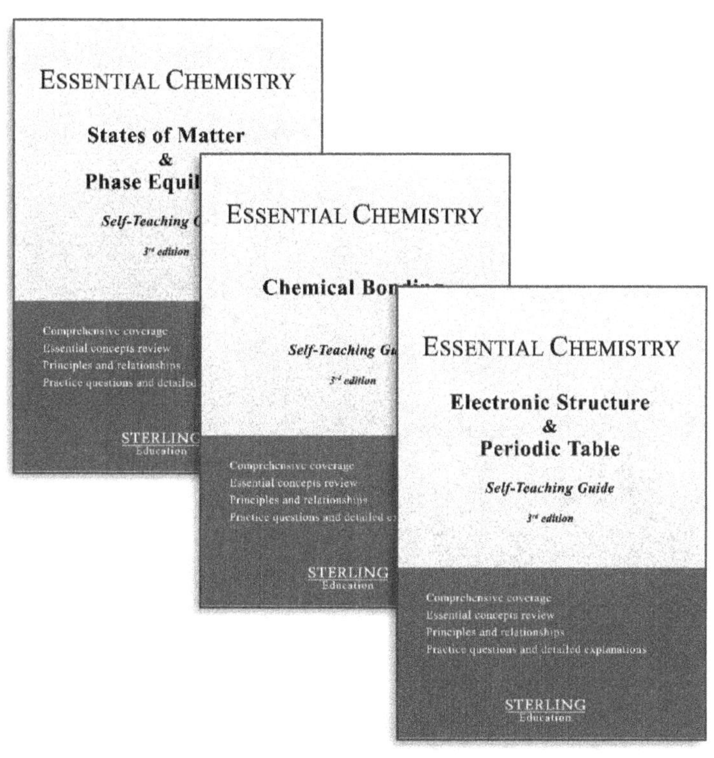

For online practice resources visit

https://www.sterling-prep.com

If you benefited from this book, please leave a review on Amazon so others
can learn from your input. Reviews help us understand our customers'
needs and experiences while keeping our commitment to quality.